7th EEIGM International Conference on Advanced Materials Research 2013

IOP Conference Series: Materials Science and Engineering
Volume 48

Lulea, Sweden
21 - 22 March 2013

ISBN: 978-1-63439-859-6
ISSN: 1757-8981

7th EEIGM International Conference on Advanced Materials Research 2013

IOP Conference Series: Materials Science and Engineering
Volume 48

Lulea, Sweden
21 - 22 March 2013

ISBN: 978-1-63439-859-6
ISSN: 1757-8981

IOP Institute of Physics

Printed from e-media with permission by:

Curran Associates, Inc.
57 Morehouse Lane
Red Hook, NY 12571

Some format issues inherent in the e-media version may also appear in this print version.

Printed by Curran Associates, Inc. (2015)

For permission requests, please contact the Institute of Physics
at the address below.

Institute of Physics
Dirac House, Temple Back
Bristol BS1 6BE UK

Phone: 44 1 17 929 7481
Fax: 44 1 17 920 0979

techtracking@iop.org

Additional copies of this publication are available from:

Curran Associates, Inc.
57 Morehouse Lane
Red Hook, NY 12571 USA
Phone: 845-758-0400
Fax: 845-758-2634
Email: curran@proceedings.com
Web: www.proceedings.com

Table of contents

Volume 48

7th EEIGM International Conference on Advanced Materials Research
21–22 March 2013, LTU, Luleå, Sweden

Accepted papers received: 28 October 2013
Published online: 12 December 2013

Papers

012001

Improvement of mechanical properties on metastable stainless steels by reversion heat treatments OPEN ACCESS A Mateo, A Zapata and G Fargas pg. 1

012002

Study of alpha-case depth in Ti-6Al-2Sn-4Zr-2Mo and Ti-6Al-4V OPEN ACCESS R Gaddam, B Sefer, R Pederson and M-L Antti pg. 7

012003

Mechanical response under contact loads of AlCrN-coated tool materials OPEN ACCESS J Yang, C A Botero, N Cornu, G Ramírez, A Mestra and L Llanes pg. 15

012004

Method for the evaluation of stretch blow molding simulations with free blow trials
OPEN ACCESS J Zimmer and M Stommel pg. 22

012005

Tie molecules, morphology and confinement effects in semi – crystalline poly(ethylene naphthalene-2,6-dicarboxylate) (PEN) OPEN ACCESS E Nikaj, I Royaud and L David pg. 32

012006

Influence of the mechanical stress and the filler content on the hydrostatic compression behaviour of natural rubber OPEN ACCESS J Zimmermann and M Stommel pg. 45

012007

Development of constitutive model for composites exhibiting time dependent properties
OPEN ACCESS L Pupure, R Joffe, J Varna and B Nyström pg. 55

012008

Correlation between mechanical and dielectric properties of Alfa/Wool/Polymeric hybrid fibres reinforced polyester composites OPEN ACCESS A Triki, M A Omri, M Guicha, M B Hassen, M Arous and A Kallel pg. 65

012009

Degradation of elastic properties of non-uniformly damaged composite laminates OPEN ACCESS J Varna, M S Loukil and Z Ayadi pg. 76

012010

Influence of hydrogen environment on fatigue crack growth in forged Ti-6Al-4V: fractographic analysis OPEN ACCESS R Gaddam, R Pederson, M Hörnqvist and M-L Antti pg. 87

012011

The effect of crystallographic orientation on solid metal induced embrittlement of Ti-8Al-1Mo-1V in contact with copper OPEN ACCESS P Åkerfeldt, R Pederson, M-L Antti, Y Yao and U Klement pg. 96

012012

Case study of grate-chain degradation in a Grate-Kiln process OPEN ACCESS E A A Nilsson, L Pettersson and M-L Antti pg. 106

012013

Effects of non-hydrostatic pressure on electrical resistance of bundled single-wall carbon nanotubes OPEN ACCESS M Noël, Y Volkova, M Mases, P Zelenovskiy, A Babushkin and A V Soldatov pg. 116

IOP Conf. Series: Materials Science and Engineering **48** (2013) 012001 doi:10.1088/1757-899X/48/1/012001

Improvement of mechanical properties on metastable stainless steels by reversion heat treatments

A Mateo , A Zapata and G Fargas
Department of Materials Science and Metallurgical Engineering,
Universitat Politècnica de Catalunya, Avda. Diagonal 647, 08028 Barcelona, Spain.

E-mail: antonio.manuel.mateo@upc.edu

Abstract. AISI 301LN is a metastable austenitic stainless steel that offers an excellent combination of high strength and ductility. This stainless grade is currently used in applications where severe forming operations are required, such as automotive bodies. When these metastable steels are plastically deformed at room temperature, for example by cold rolling, austenite transforms to martensite and, as a result, yield strength increases but ductility is reduced. Grain refinement is the only method that allows improving strength and ductility simultaneously. Several researchers have demonstrated that fine grain AISI 301LN can be obtained by heat treatment after cold rolling. This heat treatment is called reversion because it provokes the reversion of strain induced martensite to austenite. In the present work, sheets of AISI 301LN previously subjected to 20% of cold rolling reduction were treated and a refined grain austenitic microstructure was obtained. Mechanical properties, including fatigue limit, were determined and compared with those corresponding to the steel both before and after the cold rolling.

1. Introduction

Metastable steels are highly suitable for the fabrication of automotive components, since they have good formability and, after forming by processes like stamping, hydroforming or cold rolling, the components achieve a high strength, required to meet the safety standards regarding crash behavior. In the particular case of metastable austenitic stainless steels, the plastic deformation induces the transformation from austenite to martensite, which increases the hardening coefficient and gives the material an extraordinary energy absorption capacity [1]. Due to these outstanding characteristics, AISI 301 grade is widely used for construction of lightweight train structures [2], bus carriages [3], honeycomb structures and it is been introducing in the automotive industry for impact supporting frames such as the B-pillar [4].

Furthermore, various authors have shown that it is possible to simultaneously improve strength and ductility of metastable austenitic steels by means of heat treatments [1,5-7]. These treatments are applied to previously cold worked steels with the purpose of achieving recrystallization of the predeformed austenite and simultaneous reversion of the strain induced martensite, so that an austenitic microstructure of ultrafine [6] or even nanocrystalline [7] grain size can be obtained.

Based on the state of the art above described, the present work deals first with the determination of optimal conditions for reversion treatments on a stainless steel AISI 301LN after a 20% thickness reduction by cold rolling. Once these reversion treatment conditions were established, tensile properties, hardness and fatigue limits, were measured for the fine grain steel accordingly obtained.

2. Materials and Experimental Procedure

2.1. Material

The stainless steel AISI 301LN, with the designation EN 1.4318 according to the European standard, was provided by OCAS NV, Arcelor-Mittal R&D Industry Gent (Belgium) as 1.5 mm thick sheets. Its chemical composition is presented in table 1.

Table 1. Chemical composition of the steel AISI 301LN (wt %).

C	Cr	Ni	Mo	Mn	Si	N
0.03	17.5	7.10	0.10	1.48	0.50	0.14

The material was provided in two conditions:

a) Cold rolled, with subsequent solution annealing (fast cooling from temperatures around 1050 ºC), a treatment aiming at a completely austenitic microstructure free of second phase precipitations.
b) As before, but subjected to a final cold rolling with a thickness reduction of 20%, process resulting in a microstructure consisting of an austenitic matrix besides deformation induced martensite plates.

2.2. Reversion treatments

In agreement with literature [1,5-7] as well as with previous research carried out by the authors of this paper [8], the temperature was chosen in the range between 600 and 850 ºC. At lower temperatures martensite reversion cannot be totally achieved, while at higher temperatures the recrystallized austenite grains may grow too fast. The holding times were 1, 5 and 10 minutes, with exception of the test at the highest temperature, where the time was shortened to 30 s and 1 min.

Tensile tests were carried out with samples treated in an electric resistance oven. Samples were introduced into the oven once the specified holding temperature had stabilized, while the quenching was carried out by fast immersion into water.

2.3. Microstructural characterization

In case of metastable steels grinding and mechanical polishing can induce martensite formation at the sample surface. In order to avoid this effect, electropolishing in a nitric acid solution was performed. It was not possible to reveal simultaneously austenite and martensite. Thus, the same etching solution was used for electrolytic etching to visualize austenite grains, while martensite was revealed by immersion in Beraha solution [9].

Since micrographic quantification of the martensite fraction induced by cold rolling is little reliable, X-ray diffraction was applied, by using the copper radiation, and the phase fractions were calculated with the RIR method (*Reference Intensity Ratio*), as explained in detail in reference [10].

2.4. Mechanical properties

Vickers hardness testing with a load of 100 g was performed. A minimum of six indentations for each material condition were done. Subsequently, tensile tests were carried out with three specimens per condition. All tensile specimens were oriented longitudinally, that is with the tensile axis parallel to the cold rolling direction. Test rate was of 3 mm/min.

Considering the tensile strength, a testing procedure based on the staircase (or up-and-down) method was applied to determine the fatigue limits of AISI 301LN in the different material conditions [11]. The first step consisted in applying a maximum load (σ_{max}) of around 55% of the ultimate tensile strength of the corresponding steel condition. If the specimen failed before infinite life (that is $2x10^6$

cycles), the next specimen had to be tested at a 5% lower stress level. If the specimen did not fail, the next test was run at a higher stress level with the same specimen, and so on until fracture. Despite it is recommended to run the staircase with at least 15 specimens, a more reduced number (at least 8 per condition) were used for this research. Tests were conducted in a resonant testing machine at frequencies around 150 Hz with a stress ratio ($R= \sigma_{min}/\sigma_{max}$) of 0.1. Fatigue specimens were cut by laser from the sheets. Laser cutting is widely used in the metal industry, particularly when components of high strength steel sheets are produced. On the other hand, the roughness of cut edges produced by laser differs from that obtained by mechanical cutting, and this may influence the fatigue performance. Therefore, fatigue specimens were carefully polished up to R_a values under 1 μm before the tests.

3. Results and Discussion

3.1. Microstructural characterization

Figure 1 shows a micrograph of the sheet surface in annealed condition. The microstructure is almost completely austenitic, with equiaxial grains of 11.7 ± 4 μm mean size and abundant presence of twins. After cold rolling the microstructure remained similar with respect to grain size, but the martensite volume fraction determined from X-ray diffractograms was 28%.

Figure 1. Micrograph of the AISI 301LN steel in annealed condition.

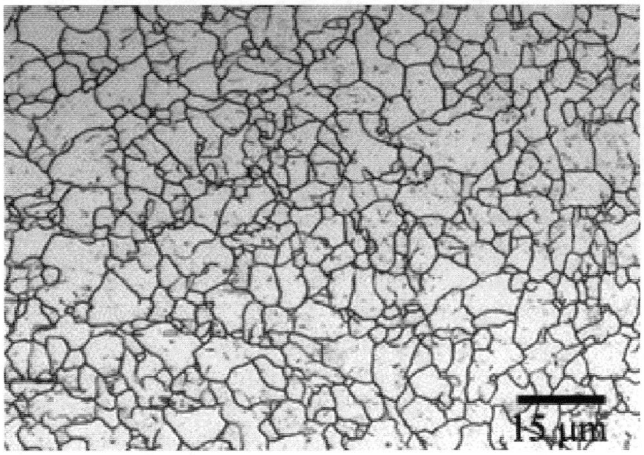

Figure 2. Micrograph of the AISI 301LN after a reversion treatment of 10 minutes at 750 ºC.

The respective values of the achieved grain sizes after reversion heat treatments are given in table 2. From these results, it is evident that all the treatments produce a decrease in grain size. However, the best grain refining was achieved at 750 ºC and a holding time of 10 minutes. Therefore, these were the heat treatment conditions selected to perform the mechanical characterization. At this temperature there is no risk of growing for the recrystallized austenitic grain, since Di Schino et al. [5] showed that this event will take place only at temperatures above 900 ºC.

Figure 2 shows the microstructure after reversion at the optimal conditions and allows comparing it with the annealed steel at the same magnification (Figure 1). Grain size distribution indicates a predominance of ultrafine grains (size between 0.5 and 1.5 μm), but coexisting with bigger grains (probably austenitic grains which did not recrystallize during the reversion treatment). This heterogeneous distribution is reflected in the dispersion level, which is almost of the same value as the average grain size: 2.9 ± 2.8 μm.

Table 2. Average austenitic grain sizes (in μm) resulting from reversion heat treatments.

t (min) / T (ºC)	600	700	750	850
0.5	-	-	-	4.6
1	7.8	7.8	4.0	4.9
5	7.7	8.7	3.7	-
10	7.5	7.5	2.9	-

Concerning martensite dissolution, neither α'-martensite phase nor ε–martensite were observed after the reversion treatment. Figure 3 compares the XRD spectrum corresponding to the cold rolled condition, where α' peaks are present, with the one for the steel after reversion where only austenite peaks appear.

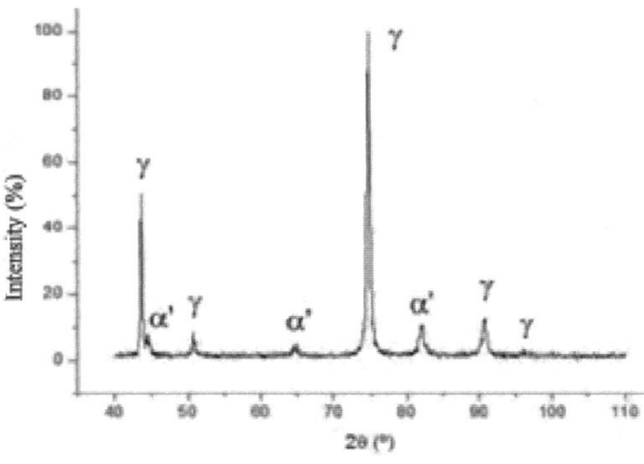

a)

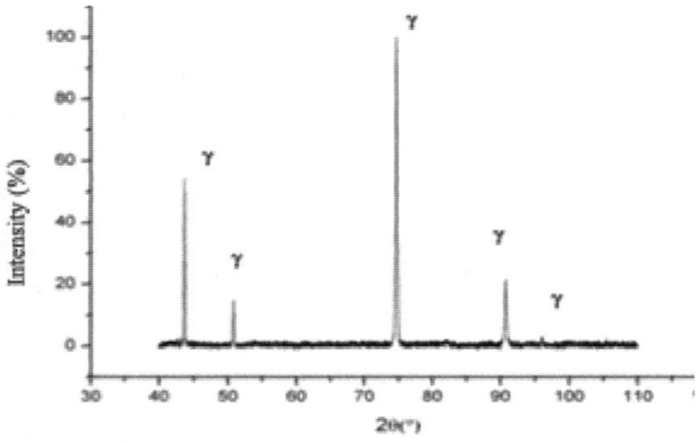

b)

Figure 3. X-ray diffraction spectra of AISI 301LN steel: a) Cold rolled; b) After reversion treatment.

3.2. Mechanical characterization

Table 3 compares the values of hardness, yield strength, ultimate tensile strength and elongation for AISI 301LN in the two initial conditions, as well as after reversion treatment of the cold rolled material.

Values in table 3 show the excellent combination of static properties exhibited by the 301LN steel after the reversion treatment. Especially outstanding is its yield strength which more than doubles the value corresponding to the annealed condition. On the other hand, the hardness and grain size values of the completely austenitic microstructures, i.e. annealed and reversion conditions, are in good agreement with Di Schino et al. [5], who reported the hardness of AISI 301 steels to depend on the

IOP Conf. Series: Materials Science and Engineering **48** (2013) 012001 doi:10.1088/1757-899X/48/1/012001

grain size according to the Hall-Petch relationship for grains larger than 3 μm. For finer grains a saturation value around 330 HV0.1 was found.

Table 4 gives the fatigue limits obtained after statistical calculations based on the staircase tests performed. Results are indicated both in terms of stress amplitude, as well as maximum applied stress. After reversion the steel exhibits an improvement of almost 100 MPa (in terms of $\Delta\sigma/2$), that represents a degree of 36%, as compared to the annealed condition. This substantial improvement of fatigue resistance by grain size refinement is consistent with earlier investigations [12]. The ratios between maximum stress in fatigue to yield stress and also to ultimate tensile strength are indicated in table 4 as well. Cold rolled and reversion conditions exhibit similar ratios, whereas annealed steel has an elevated ratio when yield stress is considered, but low for the relationship $\sigma_{max}/\sigma_{uts}$. This fact is associated with the very high strain hardening coefficient *n* in the annealed condition, which has an average value of 0.55, but reaches instantaneous values close to 0.80, as determined by the authors in [13].

Table 3. Hardness and tensile properties of AISI 301LN in the different conditions.

Condition	Hardness (HV0.1)	σ_{ys} (MPa)	σ_{uts} (MPa)	Elongation (%)
Annealed	246 ± 12	360	902	40
Cold rolled	408 ± 30	926	1113	24
Reversion	314 ± 11	749	1010	33

Table 4. Fatigue limit values of AISI 301LN in the different conditions.

Condition	Fatigue limit, $\Delta\sigma/2$ (MPa)	Fatigue limit, σ_{max} (MPa)	σ_{max}/σ_{ys} (%)	$\sigma_{max}/\sigma_{uts}$ (%)
Annealed	256 ± 25	570 ± 56	1.58	0.63
Cold rolled	403 ± 43	895 ± 95	0.97	0.80
Reversion	349 ± 25	775 ± 56	1.03	0.77

4. Conclusions

The results of this work show the viability of reversion heat treatments as a method to achieve a grain refinement in metastable stainless steels. In the present case, cold rolled AISI 301LN steel with a martensite fraction of 28% and an initial austenitic grain size of 12 μm was studied. The first part of the research pointed out that the optimal reversion treatment conditions were 10 minutes at 750 °C. This reversion treatment led to an austenitic mean grain size below 3 μm, although with a very heterogeneous distribution, and also to a complete disappearance of the martensite. The second part of the study proved that this fine grain steel has an excellent combination of mechanical properties, featuring a yield strength which doubles that of the annealed condition, and also a fatigue limit improvement of 36%.

References

[1] Eskandari M, Najafizadeh A, Kermaunpur A and Karimi M 2009 Potential application of nanocristalline 301 austenitic stainless steel in lightweight structures *Mater.&Design* **30** 3869-72

[2] Gales A, Sirén M, Säynäjäkangas J, Akdut N, van Hoecke D and Sánchez R 2007 Development of lightweight train and metro cars by using ultra high strength stainless steels *Official Publications of the European Communities, Technical Steel Research* Report EUR 22837 (Luxemburg)

[3] Kyröläinen A, Sánchez R, Santacreu P-O, Picozzin V and Gales A 2003 Stainless steels in bus contructions. *Official Publications of the European Communities, Technical Steel Research, Special and Alloy Steels,* Report EUR 20884 (Luxemburg)

[4] Siren M, de Wispelaere N, Rizzo L, Pauly T, Kosmac A, Taulavuori T, Sanchez R, Vliegen R, Van Hecke B, Säynevirt J and Hänninen H 2010 Innovative stainless steel applications in transport vehicles, *Official Publications of the European Communities, Technical Steel Research* Report EUR 24218 (Luxemburg)

[5] Di Schino A, Barteri M and Kenny J M 2002 Development of ultrafine structure by martensite reversion in stainless steel *J Mater Sci Letters* **21** 751-3.

[6] Somani M C, Juntunen P, Karjalainen L P, Misra R D K and Kyröläinen A 2009 Enhanced Mechanical Properties through Reversion in Metastable Austenitic Stainless Steels *Metall Mater Trans* A **40,3** 729-44.

[7] Ma Y, Eun Jin J and Kook Lee Y 2005 A repetitive thermomechanical process to produce nano-crystalline in a metastable austenitic steel *Scripta Mater* **52** 1311-5.

[8] Mateo A, Hernández A, Zapata A, Rodríguez Calvillo P, Fargas G, Calvo J and Casellas D 2012 Tratamientos térmicos de reversión en un acero inoxidable metaestable *XII Congreso Iberoamericano de Materiales IBEROMAT XII*, Univ de Alicante, Trabajo A-20

[9] Vander Voort G F 1999 Metallography: principles and practice *ASM International* vol. 557 (Ohio)

[10] Mateo A, Fargas G. and Zapata A 2012 Martensitic transformation during fatigue testing of an AISI 301LN stainless steel. IOP *Conf. Ser.: Mater. Sci. Eng.* **31** 011001 ed D Horwat et al.

[11] Lin S-K, Lee Y-L and Lu M-W 2001 Evaluation of the staircase and the accelerated test methods for fatigue limit distributions *Int J Fatigue* **23**, 75-83.

[12] Hamada AS, Karjalainen LP, Ventaka Surya PKC and Misra RDK 2011 Fatigue behaviour of ultrafine-grained and coarse-grained Cr-Ni austenitic stainless steels *Mat Sci Eng A* **528,** 3890-3896.

[13] Gutiérrez D, Mateo A, Rodríguez-Calvillo P, Fargas G, Lara A, Casellas D and Prado JM 2012 Conformabilidad de chapas de aceros metaestables *XII Congreso Iberoamericano de Materiales IBEROMAT XII*, Univ de Alicante, Trabajo K-4

Acknowledgements

The presented work was carried out within the scope of MAT09-14461 project, supported by the Spanish Ministry of Science and Innovation. We are grateful to *Direcció General de Recerca del Comissionat per a Universitats i Recerca de la Generalitat de Catalunya* for acknowledging CIEFMA as a Consolidated Research Group (2009SGR), to OCAS NV and to Outukumpu for providing the investigated material. C.H. Sacre and A. El Ouali (EEIGM-UPC students), I. Sapienzanskaia (CEIFMA-UPC), A. Hernández (CTM) and J. Calvo (UPC and CTM) are acknowledged for their collaboration.

IOP Conf. Series: Materials Science and Engineering **48** (2013) 012002 doi:10.1088/1757-899X/48/1/012002

Study of alpha-case depth in Ti-6Al-2Sn-4Zr-2Mo and Ti-6Al-4V

R Gaddam[1], B Sefer[1], R Pederson[1,2] and M-L Antti[1]
[1]Division of Materials Science, Luleå University of Technology, S-97187 Luleå, Sweden
[2]Research Centre, GKN Aerospace Engine Systems, S-46181 Trollhättan, Sweden

E-mail: raghuveer.gaddam@ltu.se

Abstract. At temperatures exceeding 480°C titanium alloys generally oxidises and forms a hard and brittle layer enriched with oxygen, which is called alpha case. This layer has negative effects on several mechanical properties and lowers the tensile ductility and the fatigue resistance. Therefore any alpha-case formed on titanium alloys during various manufacturing processes, such as heat treatment procedures, must be removed before the final part is mounted in an engine. In addition, long time exposure at elevated temperatures during operation of an engine could possibly also lead to formation of alpha-case on actual parts, therefore knowledge and understanding of the alpha-case formation and its effect on mechanical properties is important. Factors that contribute for growth of alpha-case are: presence of oxygen, exposure time, temperature and pressure. In the present study, isothermal oxidation experiments in air were performed on forged Ti-6Al-2Sn-4Zr-2Mo at 500°C and 593°C up to 500 hours. Similar studies were also performed on Ti-6Al-4V plate at 593°C and 700°C. Alpha-case depth for both alloys was quantified using metallography techniques and compared.

1. Introduction

The titanium alloys, Ti–6Al–2Sn–4Zr–2Mo (Ti-6242) and Ti-6Al-4V (Ti-64) are widely used for components in aeroengines because of their excellent combination of weight and strength [1-3]. The maximum service temperature for these alloys is limited to 450°C for Ti-6242 and 350°C for Ti-64 [2,3]. It is partly because of degradation of mechanical properties above these respective temperatures as at elevated temperatures (above 480°C), titanium alloys oxidises in oxygen containing environments; this results in simultaneous formation of an oxide (TiO_2) scale on the surface and an oxygen-rich layer beneath the scale, commonly referred to as alpha-case (α-case). Alpha-case is a continuous, hard, and brittle layer with higher oxygen content [1,2]. It forms because of higher solid solubility of oxygen in α-titanium (i.e.14.5 wt.% [4]) and higher affinity of titanium to absorb oxygen, which instantaneously reacts and stabilise the α phase. This brittle alpha-case degrades the mechanical properties such as tensile ductility and fatigue strength [5-10]. Therefore it is necessary to remove any alpha-case formed on parts manufactured from titanium alloys if they are subjected to high loads and/or dynamic loading conditions.

Environmental and economical requirements on future aero engines set higher demands on the efficiency and thereby pressure ratio. The increased pressure leads to higher temperatures, which could make it necessary to replace titanium alloys with nickel based super alloys in the front end of the engine. Unfortunately this increases the weight of the engine and leads to an increase of fuel consumption and is therefore undesirable. However, there is a possibility to increase the maximum working temperature of the currently used Ti-6242 alloy in compressor parts by developing a better understanding on how different mechanical properties and physical phenomena such as oxidation mechanisms are affected in the high temperature regimes. And since some aeroengine parts consists of a combination of both Ti-64 and Ti-6242. Hence both of these alloys are investigated in the current study. The objective of present study was to investigate the effect of heat treatment conditions on the depth of alpha-case in two common aerospace grade titanium alloys; Ti-6242 and Ti-64. The heat treatment conditions investigated include the temperatures and times in ambient air that are of interest for heat treatment during manufacturing and for application in an aeroengine for both of these alloys.

2. Materials and Methods

The materials investigated are Ti-6242 and Ti-64 with the chemical compositions shown in table 1. Ti-6242 is a near- α alloy that has been solution and precipitation heat treated; and was obtained in forged condition according to AMS 4976G [11]. It consists of a bi-modal microstructure with primary α and transformed β (see figure 1(a)). Ti-64 is a α+β titanium alloy that has been obtained in plate form according to AMS 4911L [12], which consists of a microstructure with equiaxed primary α and elongated α in transformed β (see figure 1(b)).

Table 1. Chemical composition (wt.%) of the materials investigated.

Material	Al	Sn	Zr	Mo	N	O	C	H	Fe	Si	V	Ti
Ti-6242	6.5	2.2	4.4	2.2	0.05	0.15	0.05	0.05	0.1	0.1	-	Bal.
Ti-64	6.75	-	-	-	0.05	0.20	0.08	0.015	0.3	-	4.5	Bal.

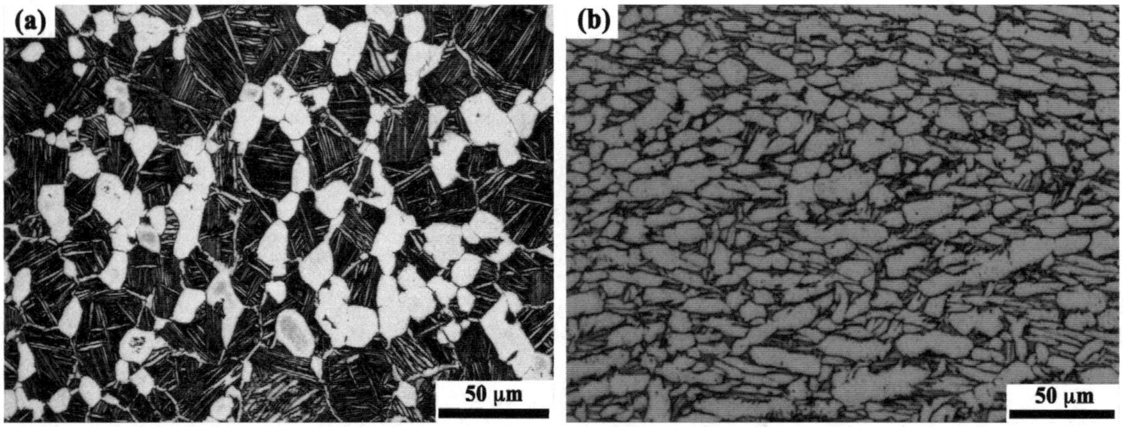

Figure 1. Optical micrographs showing the microstructure of as received materials, (a) forged Ti-6242 and (b) Ti-64 plate. In micrographs light or grey areas is α phase and dark areas is β.

2.1. Heat treatment

The samples for heat treatments were cut using electric discharged machining (EDM) from the as received material in dimensions of 10 x 5 x 10 mm. Totally 64 samples were cut out, where two samples were used for each exposure time and temperature. All the sample surfaces were metallographically polished to remove the recast layer. Thereafter, the polished samples were ultrasonically cleaned in technical acetone for approximately 15 min and rinsed with ethanol. The cleaned samples were heat treated in ambient air using a Nabertherm box furnace (N11/R) and a Nabertherm tube furnace (RHTC 80-450/15) at atmospheric pressure. The temperature inside the furnace is calibrated using a reference sample with a thermocouple welded to it and positioned at the locations inside the furnace where the samples are placed. The temperature inside the furnace was about ± 5°C of the desired temperature. The samples were placed onto an Al_2O_3 plate or crucible and then introduced into the furnaces at the desired temperature and isothermally held for selected exposure times as shown in table 2. All samples were weighed using a microbalance with an accuracy of ± 0.0001 g before and after each heat treatment. In addition, heat treatments were also performed on Ti-6242 and Ti-64 samples, with a cross section of 17 x 2.5 x 10 mm, in dry air (i.e. technical air) using a simultaneous thermal analysis instrument (STA 449C from Netschz Gmbh). The samples were isothermally held at 593°C for 200 hours.

Table 2. Heat treatment conditions.

Material	Temperature (°C)	Exposure times (hour)
Ti-6242 forged	500 593	5, 10, 50, 100, 200, 300, 400 and 500
Ti-64 plate	593 700	

2.2. Alpha-case evaluation

Heat treated samples were cut into half, parallel to the face that is placed on the crucible (10 x 5 mm^2), thereafter ultrasonically cleaned in technical acetone for 15 min and rinsed with ethanol. Cleaned samples were mounted in bakelite using BUEHLER Simplimet model 1000 mounting machine and then metallographically prepared, which involves grinding and polishing the surface up to 0.05 μm with colloidal silica using a semiautomatic BUEHLER Phoenix 4000 polishing machine. The polished samples were etched in two steps to observe the alpha-case: first, swabbing the sample surface with Kroll's reagent (a mixture of 1-3 ml HF, 2-3 ml HNO_3 and distilled water) and second, immersing in Weck's reagent (1-3 g NH_4HF_2 and distilled water) for approximately 10 sec. Alpha-case layer was observed using a Nikon Eclipse optical microscope (model MA200) and the depth of alpha-case layer was quantitatively measured using NIS elements software. Totally 60 individual measurements of alpha-case depth on each sample were made along the entire perimeter at approximately 500 μm spacing on two samples of Ti-6242 and Ti-64 at each temperature and time combination, and the average values of alpha-case depth is then reported. The hardness measurements related to the alpha-case at the surface was obtained by applying a load of 100 g using MXT- α microhardness tester from Matsuzawa with a Vickers indenter (indent size is approximately 20 – 25 μm).

3. Results and Discussion

3.1. Weight gain

Figure 2(a) presents the amount of weight gain (ΔW) per surface area (A) for Ti-6242 and Ti-64 when isothermally held at 593°C for 200 hours in dry air. It is calculated by dividing the weight gain values of the samples, measured using STA 449C, with their total surface areas. The unit is (mg/cm^2). From figure 2(a) it can be seen that the ($\Delta W/A$) increased with time and approximately followed a parabolic relationship for both the alloys. Similar behaviour for both the alloys was observed on samples that were isothermally held in a laboratory furnace at 593°C for up to 500 hours in ambient air, where the weight gain is calculated by weighing before and after the heat treatment. It was noted that in Ti-6242 samples held at 500°C for up to 500 hours the weight gain followed approximately a parabolic relationship, while deviating from parabolic below 100 hours. In contrast, Ti-64 samples held at 700°C showed deviation from parabolic relationship beyond 200 hours. The observation of weight gain in both the alloys is obtained by fitting and performing regression analysis of the weight gain data using the following equation [13]:

$$(\Delta W/A) = Kt^{1/n} \tag{1}$$

where K is the rate constant and n is the reaction index. At a constant temperature, the weight gain is assumed to be linear if n=1, parabolic if n=2. The weight gain in the Ti-6242 alloy at the tested temperatures is consistent with other investigators [8, 14]. On the other hand, the Ti-64 alloy that was held at 700°C for up to 500 hours, indicated a transition from parabolic to linear weight gain at about 200 hours. This confirms with the results found by others [15-17]. In figure 2(a) it can be seen that weight gain is much higher in Ti-64 compared to Ti-6242 at the constant temperature and time. This could be due to the thicker oxide layer (≈ 5 μm) in Ti-64 than in Ti-6242 (< 1 μm), see figure 2(b) and figure 2(c).

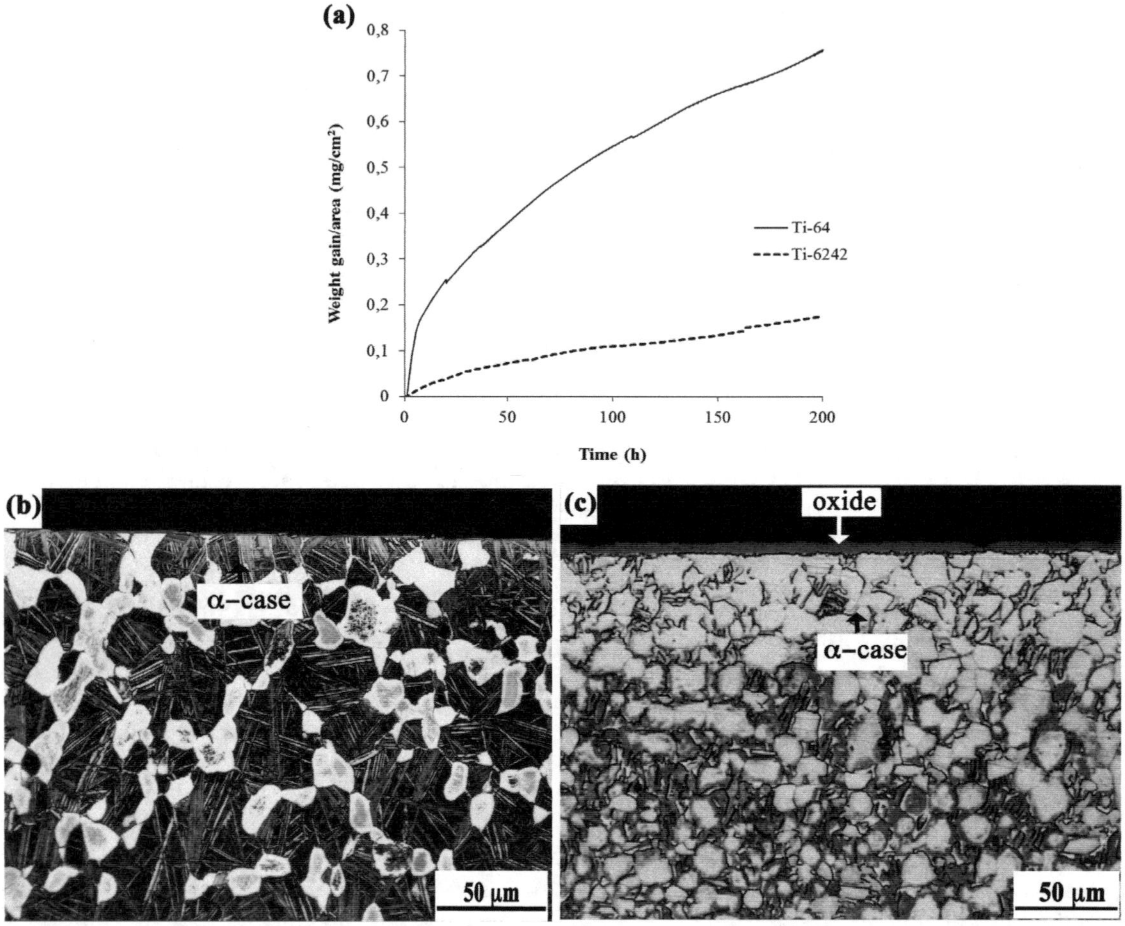

Figure. 2 (a) Plot showing the weight gain per unit area (mg/cm^2) in Ti-6242 and Ti-64 held at 593°C for 200 hours in dry air, (b,c) optical micrographs of Ti-6242 and Ti-64, showing the oxide layer and alpha case.

3.2. Alpha-case evaluation

Figure 3 shows the representative optical micrographs of alpha-case (white layer) in Ti-6242 and Ti-64 after long exposure time (500 hours) at 593°C. It can be seen that alpha-case is formed beneath the surface simultaneously with the oxide (see Figure 3(b)). Figure 4 shows the plot of alpha-case depth versus exposure time for Ti-6242 and Ti-64 at the tested temperatures. Here, alpha-case is measured quantitatively using the optical micrographs. From figure 4, it can be seen that depth of alpha-case is increased with temperature and time in both the alloys and mainly follows a parabolic relationship at the selected heat treatment conditions. The maximum depth of alpha-case formed in Ti-6242 is approximately 30 μm when exposed at 593°C for 500 hours, and 10 μm at 500°C after 500 hours (see figure 4). The depth of alpha-case in Ti-64 when exposed to 593°C for 500 hours was about 30 μm. In addition, it is observed that a thick alpha-case layer of approximately 200 μm is formed in Ti-64 when exposed at 700°C for 500 hours.

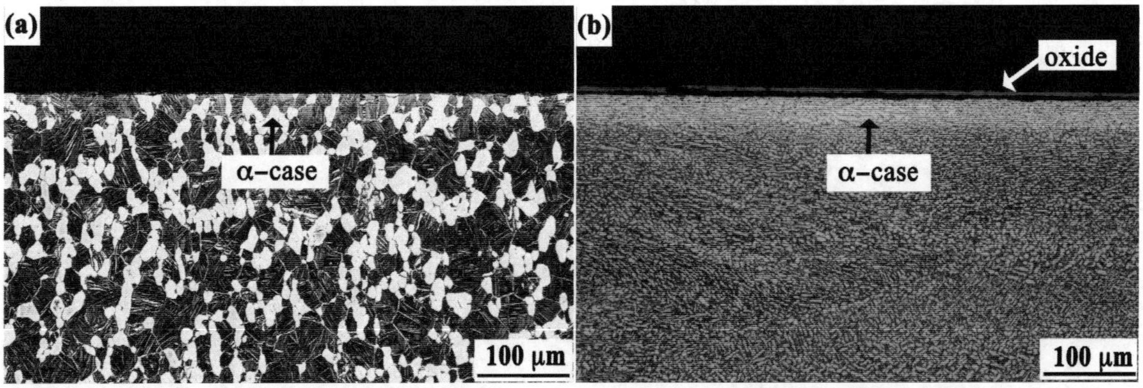

Figure 3. Optical micrographs showing alpha-case (white layer) in (a) Ti-6242 and (b) Ti-64 heated at 593°C up to 500 hours.

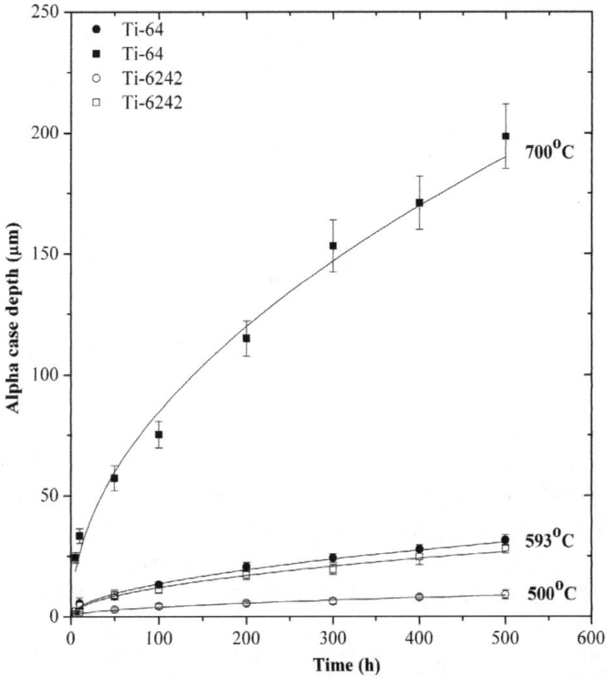

Figure 4. Variation of alpha-case depth in Ti-6242 and Ti-64 as a function of time at different temperatures. Empty data points correspond to Ti-6242 and filled data points are for Ti-64.

From figure 4, it can be seen that the growth of alpha-case in both Ti-6242 and Ti-64 alloys exposed at 593°C for 500 hours follows a parabolic relationship, which can be related to the bulk diffusion. The approximate solution that describe Fick's second law of diffusion is given by $x=\sqrt{(Dt)}$, where x is alpha-case depth, D is the diffusion coefficient (m^2/s), and t is exposure time. Figure 5 shows a log-log plot of alpha-case depth versus exposure time. It shows a linear relationship with R-values 0.995. The exponents from the linear fits are nearly equal to 0.5. Hence, the values for D are calculated from the slope obtained from the best-fit lines. The D calculated here is found to be in the order of 10^{-16} m^2/s for Ti-6242 and Ti-64. From the present study, it can be seen that there is no significant change in the depth of alpha-case and the diffusion of oxygen in Ti-6242 and Ti-64 when exposed at 593°C up to 500 hours. This shows that there might be no substantial influence of the difference in chemical composition and the microstructure between the two alloys on alpha-case depth. Similar observations were also noted for both the alloys when exposed at 704°C for 24 hours in air [18].

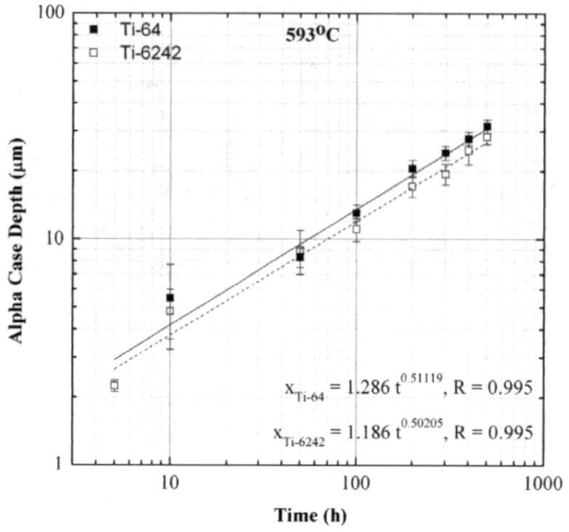

Figure 5. Plot showing the alpha-case depth vs exposure time for Ti-6242 (empty boxes) and Ti-64 (filled boxes) at 593°C up to 500 hours (log- log scale).

Alpha-case is commonly referred to a region enriched with oxygen. It is known that oxygen stabilizes the α phase and increases the strength of titanium by solid solution strengthening [2-3]. Therefore the alpha-case layer is harder than the bulk. Table 3 shows the results obtained from the microhardness measurements on Ti-6242 that are isothermally treated at 593°C; Ti-64 at 593 and 700°C for up to 500 hours. It was found that the hardness values in the alpha-case are higher in magnitude than in the bulk. Figure 6 shows the variation of hardness on the sample held at 700°C for 500 hours from the surface to the bulk. It can be seen that hardness values are gradually decreasing from the surface into the material to approximately 250 μm. From optical measurements on the samples held at 700°C for 500 hours, the depth of alpha-case is approximately measured to be 200 μm (see figure 4 and figure 6). The hardness values decrease with decreasing oxygen content. The difference in depth of alpha-case measured in optical microscope (OM) and indicated by the plateau in the hardness measurements suggests that there are uncertainties in identifying the border between the alpha-case layer and the bulk material in optical microscope after etching.

Table 3. Average hardness values (HV) for Ti-6242 and Ti-64 exposed at different temperatures.

	Ti-6242	Ti-64	
	593°C	593°C	700°C
Alpha case	511 ± 40	412 ± 37	630 ± 34
Bulk	362 ± 23	332 ± 10	300 ± 2.88

IOP Conf. Series: Materials Science and Engineering **48** (2013) 012002 doi:10.1088/1757-899X/48/1/012002

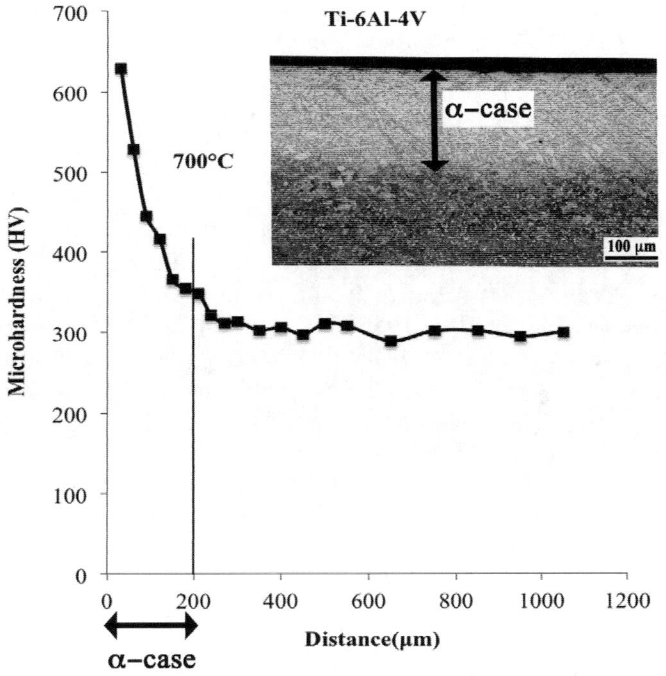

Figure 6. Variation of hardness from the surface to the bulk of the Ti-64 sample exposed to 700°C for 500 hours. An optical micrograph of the corresponding sample showing the alpha-case is also included (the white layer in the micrograph).

4. Conclusions

In the present work, isothermal heat treatments were performed on the Ti-6242 and Ti-64 titanium alloys in ambient air and at atmospheric pressure in order to study the depth of alpha-case. The conclusions are as follows:

1. Alpha-case depth in both alloys increases with temperature and time. Alpha-case growth mainly follows a parabolic relationship at 500 and 593°C in Ti-6242, and at 593°C and 700°C in Ti-64.

2. At 593°C after 500 hours Ti-6242 and Ti-64 have similar values of alpha-case depth (≈ 30 μm).

3. The alpha-case depth in Ti-6242 is 10 μm after 500 hours at 500°C.

4. The alpha-case depth in Ti-64 is 200 μm after 500 hours at 700°C.

5. The microhardness is higher in the alpha-case and decreases into the bulk.

6. A thicker oxide layer forms on Ti-64 then on Ti-6242 when isothermally held at 593°C for 200 hours. This results in a higher weight gain/area for the Ti-64 alloy. The weight gain followed a parabolic relationship.

5. References

[1] Lutjering G and Williams J C 2007 *Titanium* 2nd ed (Berlin, Springer-Verlag)
[2] Leyens C and Peters M 2003 *Titanium and titanium alloys* (Wiley VCH Verlag GmbH, Weinheim) 187
[3] Boyer R R 1996 *Mater. Sci. Eng.* A **213(1-2)** 103
[4] Murray J L and Wriedt H A 1990 Binary alloy phase diagrams Massalski T B ed. (ASM International, Materials Park, Ohio) 2924
[5] Shamblen C E and Redden T K 1968 *Proc. of The Science, Technology, and Application of Titanium*, Jaffee R I and Promisel N E eds, (Pergamon Press, Oxford, United Kingdom) 199
[6] Ruppen J A and McEvily A J 1979 *Fatigue Engg. Mater. Struct.* **2** 63
[7] Bendersky L and Rosen A 1980 *Engg. Fract. Mech.* **13** 111
[8] Shenoy R N, Unnam J and Clark R K 1986 *Oxid. Met.* **26** 105
[9] Fukai H, Iizumi H, Minakawa K-N and Ouchi C 2005 *ISIJ Inter.* **45(1)** 133
[10] Chan K S, Koike M, Johnson B W and Okabe T 2008 *Metall. Mater. Trans.* **39A** 171
[11] AMS 4976 G 2008 Titanium alloy, forgings 6.0Al-2.0Sn-4.0Zr-2.0Mo solution and precipitation heat treated (SAE-AMS/MAM)
[12] AMS 4911 L 2008 Titanium alloy, sheet, strip, and plate 6Al-4V annealed (SAE-AMS/MAM)
[13] Kofstad P 1988 *High temperature corrosion* (Elseiver Applied Science, Essex)
[14] McReynolds K S and Tamirisakandala S 2011 *Metall. Mater. Trans.* A **42A** 1732
[15] Guleryuz H and Cimenoglu H 2009 *J. Alloys Compd.* **472** 241
[16] Frangini S, Mignone A and De Riccardis F 1994 *J. Mater. Sci.* **29** 714
[17] Du H L, Datta P K, Lewis D B and Burnell Gray J S 1994 *Corros. Sci.* **36(4)** 631
[18] Unnam J, Shenoy R N and Clark R K 1984 Effect of alloy chemistry and exposure conditions on the oxidation of titanium Report No. NASA TM-86295 (NASA Technical Memorandum)

Acknowledgments

The authors kindly acknowledge Graduate School of Space Technology and National Aviation Engineering Research Programme (NFFP) for funding the present work through research collaboration with GKN Aerospace Engine Systems, Sweden. We would also like to acknowledge Francois Mattera for assisting in the experiments.

IOP Conf. Series: Materials Science and Engineering **48** (2013) 012003 doi:10.1088/1757-899X/48/1/012003

Mechanical response under contact loads of AlCrN-coated tool materials

J Yang[1,2], C A Botero[1], N Cornu[1], G Ramírez[1], A Mestra[1,2] and L Llanes[1,2]

[1]Departamento de Ciencia de los Materiales e Ingeniería Metalúrgica, ETSEIB, Universitat Politècnica de Catalunya, Avda. Diagonal 647, 08028 Barcelona, Spain

[2]CRNE, Campus Diagonal Sud, Edificio C', Universitat Politècnica de Catalunya, Barcelona 08028, Spain

Email: Jing.yang1@upc.edu

Abstract. The mechanical behavior under contact loading of systems consisting of PVD AlCrN film deposited onto two distinct hard substrates - cemented carbides and tool steel is studied by means of indentation testing techniques, under monotonic and cyclic condition. Experimental work includes assessment of critical applied loads for emergence of circular cracks at the coating surface, as well as evaluation of both surface and subsurface damage evolution. Results indicate that both coated systems are susceptible to mechanical degradation associated with repetitive contact load. Furthermore, significant differences on contact fatigue behavior between the two studied coated systems are evidenced under consideration of cracking evolution at top surface and penetration towards the substrate. In this regard, the intrinsic mechanical properties of the substrate are pointed out as key feature for rationalizing the experimental findings.

1. Introduction

Surface coatings are widely applied to improve the lifetime and performance of a large variety of tool materials because of their attractive properties, such as high hardness, good wear and chemical stability. Recently, a new ternary nitride AlCrN obtained by physical vapor deposition (PVD) techniques with higher percentage of aluminum has become the subject of ever-increasing interest due to its excellent tribomechanical properties particularly under high temperature [1,2].

Considering the complex service conditions to which coated tool systems are subjected (abrasive and adhesive wear, impact, repetitive mechanical contact, etc.) continuous research has been conducted in order to investigate the mechanical response of these coated systems and optimize the design process. In general, extensive literature exists on tribomechanical response primarily concentrated on evaluation of hardness, scratch resistance, friction coefficient and wear behavior as a function of either film nature or architecture-single versus multilayer [3–6]. However, experimental data on the mechanical performance of coated tools subjected to repetitive contact loading are rather scarce. On the other hand, experimental and analytical approaches using spherical indenters to deliver stresses over a small area of specimen surface, i.e. testing protocols based on the Hertzian theory, have been proven to be successful on the assessment of contact damage in bulk polycrystalline ceramics [7–9], layered structures, and more recently in cemented carbides and tool steels [10,11]. Within this framework, it is the aim of this study to assess the contact mechanical response and the corresponding damage mechanisms, under both monotonic and cyclic loading conditions, of an AlCrN film deposited onto two different hard substrates- cemented carbides and tool steels. In doing so, experimental protocols based on spherical indentation testing techniques are conducted.

2. Experimental procedure

Two different substrates were used in this study: a commercial fine-grained WC-10%wt Co cemented carbide grade, and a powder metallurgy processed tool steel developed by UDDEHOLM and commercialized as VANADIS. Elastic modulus and hardness for both base materials, were approximately 540 GPa and 14.5 GPa for cemented carbide [12], and 230 GPa and 7.4 GPa for the tool steel, respectively.

Aluminum chromium nitride coating was deposited on both substrates following cathodic arc PVD process. In all the cases, dense and uniform coatings with a crystalline fine-grained structure were attained (figure 1). The coatings thickness was found to be similar: 2.8 μm and 3.1 μm for the coatings deposited on the cemented carbide and the tool steel, respectively. Intrinsic hardness for coatings on both substrates was determined by means of a nanoindentation testing device (MTS Nanoindenter XP) equipped with a continuous stiffness modulus. For the two cases: coated cemented carbide and coated tool steel, hardness-penetration depth curve showed a plateau around 35 GPa and 32 GPa respectively, associated with the intrinsic hardness of the coating. Adhesion resistance was determined by means of scratch testing (Revetest, CSM Instruments) on the basis of the critical normal load related to initial coating detachment. Critical loads in both cases were higher than 30N, as given by values of 84N and 51N onto cemented carbide and tool steel respectively. Load levels above 30N are generally described as sufficient in scratch testing with a Rockwell C diamond tip for tooling applications [13].

Mechanical contact response of the coated system was investigated by means of spherical indentation. Hertzian tests were conducted in a servohydraulic testing machine (Instron 8511) by using a hardmetal spherical indenter with a curvature radius of 1.25 mm. They were focused on the failure scenario at the

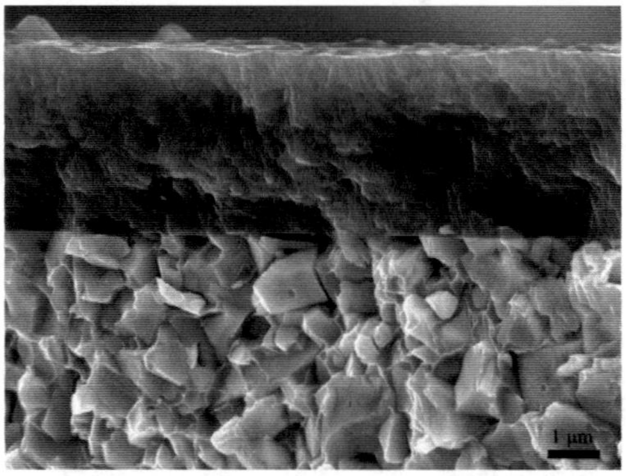

Figure 1. SEM micrograph of a fractured cross-section of the deposited AlCrN film showing the columnar character of the coating.

surface of the coated systems (i.e. circumferential cracks, cohesive failures, etc.), under monotonic and cyclic loading. In doing so, the first stage on the experimental protocol followed was the determination of the critical load for circular crack emergence at the coating surface under monotonic loading, P_c. Once it was assessed, fatigue testing was carried out at discrete maximum cyclic loading levels, P_{max}, corresponding to values equal to or close to the monotonic critical load. The cyclic loading was imposed by means of a sinusoidal waveform at a frequency of 10 Hz and corresponding ratio of 0.1 for 10^5 cycles. Subsurface damage features associated with spherical indentation were inspected by means of FIB (Zeiss Neon 40) cross-section observation.

3. Results and discussion

3.1 Monotonic spherical indentation

As applied load increases, irreversible deformation of the coated system through residual surface traces was discerned. At relatively high load levels, first signs of damage were observed at the edge of the corresponding residual imprints in terms of circumferential cracks (figure 2). The critical load for discerning film rupture was about 50% higher (1500 N as compared to 1000 N) for the WC-Co substrate with respect to the tool steel one. From a simple optical observation of the indentation imprints, it was clear that plastic yielding of the coated sets was required for the subsequent cracking in the film, in agreement with previous studies by other authors (e.g. Ref[14]) The above monotonic critical loads were then used as baseline reference for comparison purposes with the ones determined under cyclic loading conditions.

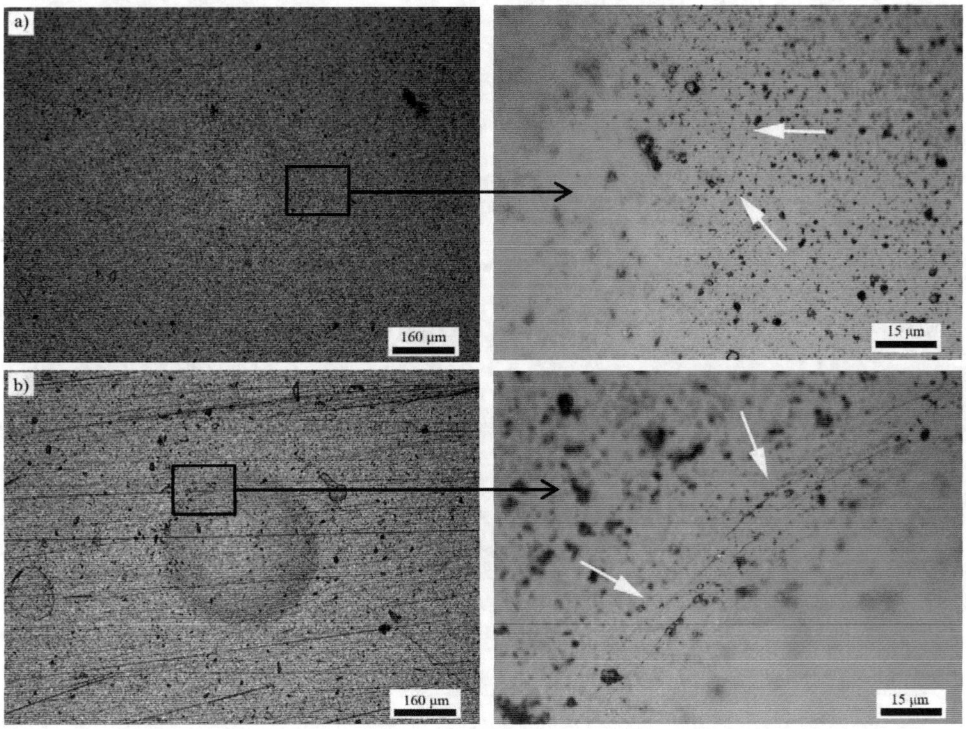

Figure 2. Circumferential cracks (indicated by white arrows) discerned under monotonic loading at the surface of the AlCrN deposited onto: a) WC-Co, and b) tool steels. (Lines on left figure 2b are inherited from the coating's original surface)

3.2 Cyclic Spherical indentation

Cyclic indentation tests were conducted attempting to evaluate the damage evolution on the studied systems as applied cyclic load increases. In both cases, experimental result showed that under cyclic loading, coated substrates exhibited accumulated damage as well as evidence of other failure mechanisms which were not observed under monotonic loading. This was clearly indicated in figure 3. Regarding the images, it should be mentioned that they correspond to insets from residual cyclic indentation imprints, as those given at the right-hand side ones in figure 2.

For the WC-Co+AlCrN system, it can be seen that under cyclic conditions, at 1000 N, a load lower than the corresponding critical monotonic load, circular cracks and small cohesive failures were observed. After applying 1500 N under fatigue condition, damage becomes more severe. For the Tool Steel+AlCrN system, cohesive failure and circular cracks were visible at 1000 N, the same value as the monotonic critical one of this system. Moreover, accumulated damage effect was also discerned as load was increased from 1000 N to 1500 N. Thus, the result confirmed that these two systems are fatigue sensitive.

Regarding the cohesive failure, it was suggested that droplets on top surface of the coating are responsible for the cohesive failure. They are typical microstructural heterogeneities in films deposited by cathodic evaporation. From a mechanical perspective, they are weakly bonded to the bulk coating and

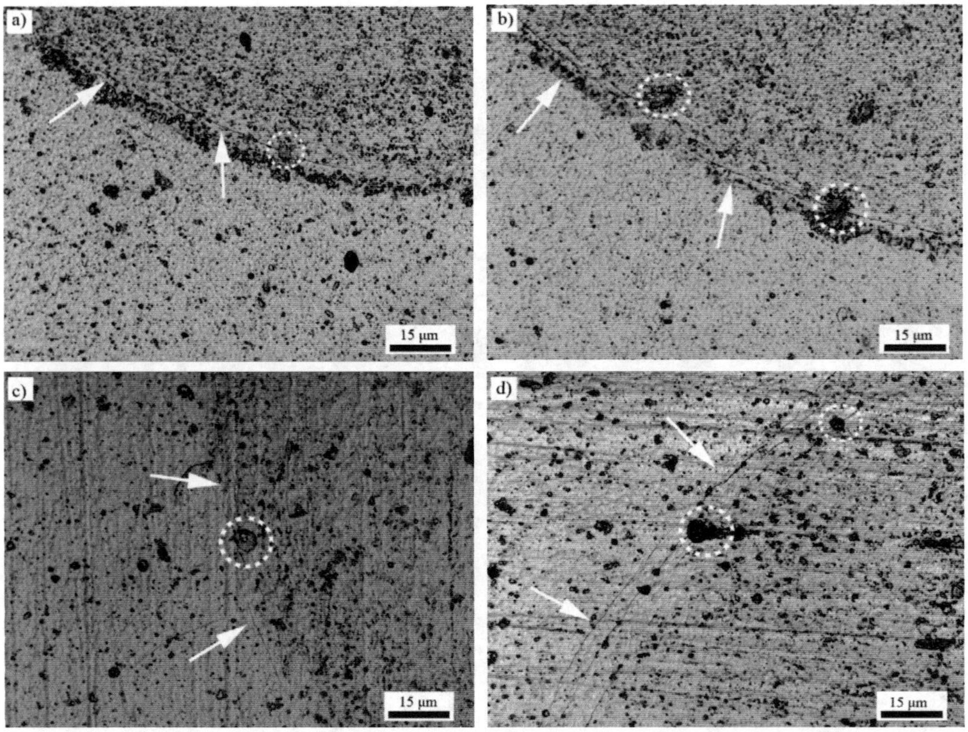

Figure 3. Damage evolution under cyclic spherical indentation on the studied systems: WC-Co+AlCrN for applied P_{max} of a) 1000 N and b) 1500 N; as well as Tool Steel+AlCrN for applied P_{max} of c) 1000 N and d) 1500 N. Circular cracks and cohesive failures are indicated by white arrows and dotted circles, respectively.

prompted to drop out from the surface under tribomechanical service conditions. Following the above ideas, cohesive failure maybe explained on the basis of progressive degradation under cyclic loading of the flawed interfaces between the droplets and bulk coatings [11].

As indicated before, the above experimental facts definitely revealed the existence of real fatigue sensitivity for both coated systems. However, it should be highlighted here that the two systems showed different damage modes under the same condition. For instance, at 1000 N during 10^5 cycles, Tool Steel+AlCrN system presented more successive multi-cracks than the WC-Co+AlCrN system. Similar trend was even intensified at 1500 N.

Considering that for both systems, coating intrinsic properties (hardness, toughness and residual stresses) were the same, relative mechanical response as a function of the substrate should be expected on the basis of the different contact strength under both monotonic and cyclic loading, in terms of the damage scenario. In some studies by other authors [15,16], it was postulated that for hard coating exhibiting columnar structure, sliding along intercolumnar cracks is the principle deformation mechanism in the coating. Thus, the system response is a combination of an elastic film that undergoes shear cracking at a critical stress and an underlying elastic–plastic substrate. Then it seems rational to speculate that WC-Co with higher stiffness and hardness than the tool steel implies less deformation, which finally renders the upward coating easier to follow and results in less shear cracks.

3.3 Subsurface indentation damage

Spherical indentation tests are especially attractive because, different from more conventional techniques involving sharp-like indenters, they allow to monitor damage evolution within an otherwise uncracked microstructure as a function of increasing either applied load (monotonic tests) or number of cycles (cyclic tests). This is particularly interesting in the case of coated systems due to the fact that here it is important to evaluate not only the intrinsic competition between deformation and fracture mechanisms but also where they are first developing, i.e. within the coating or the substrate, or even at the interface. Within this context, the damage resulting from contact loading was inspected by focused ion beam (FIB) cross section observation (figure 4).

In figure 4, it can be seen that for the WC-Co+AlCrN system, damage within the coating starts by circular crack nucleation at the periphery of the contact and at the coating surface, and subsequently advances through the thin film up to the interface, finally leading to substrate cracking either through carbides or along the metallic binder surrounding the ceramic particles. This result was consistent with previous findings of this research group [10] on degradation of TiN coated cemented carbides under cyclic loads. Abdul-Baqi and Van der Giessen [17] studied the cracking of hard coating during indentation using the finite element method. They support the idea that primary potential locations for the initiation of coating cracks are the coating surface close to the contact edge and the coating side of the interface in the contact region, where high values of tensile radial stress are found. Circumferential cracks are found to initiate from the coating surface and to propagate towards the interface. On the other hand, for the Tool Steel+AlCrN system, findings were not the same as the previous system. Cracks were observed to grow and stop at the interface rather than propagating into the substrate. Considering this special feature, it is clear the relevant role played by the tougher character of the tool steel substrate. Lawn [9] indicated that from the standpoint of substrate design, the principal requirement is to maximize hardness in order to prevent crack nucleation. But on the other hand, once cracks do initiate, softer (and generally tougher) substrates are better suited to arrest or inhibit any penetrant cracks within the sublayer through plastic deformation, so that design once more depends on whether crack prevention or crack containment is the more pressing goal.

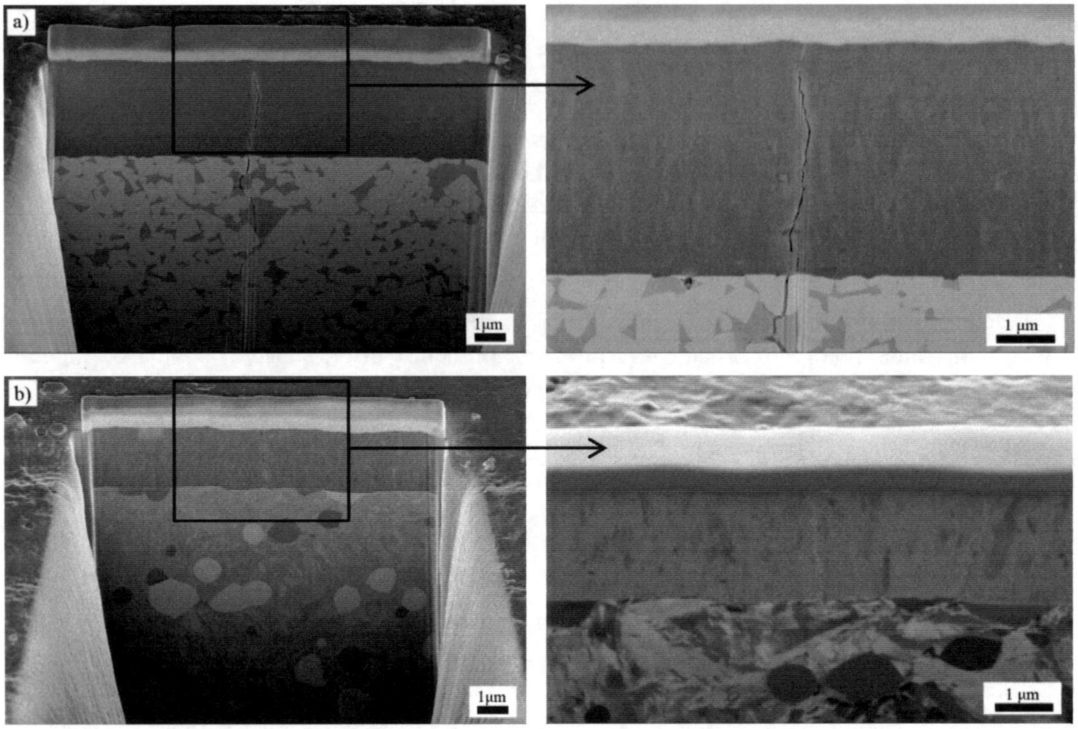

Figure 4. Typical contact damage scenario for the coated systems studied, as assessed from FIB cross-section observation: a) WC-Co+AlCrN; and b) Tool Steel+AlCrN.

Here, it should be noted that, (1) the absence of any intermediate interfacial delamination stage (at substrate/coating level) for both systems, and (2) the role of coating circular fissures as precursors of crack nucleation in the underlying brittle hardmetal. Both experimental facts are quite significant, within the context of the investigation conducted here, because they sustain the choice of early circular cracking at the coating as the critical damage event, mainly because it induces a loss of mechanical integrity of the hardmetal substrate before any adhesive failure takes place. Moreover, they also point out the relevant influence of the substrate nature on the contact response of the coated system, on the basis that such final damage evolution stage (substrate cracking) is not evidenced for the tougher substrate.

4. Conclusion

Based on the experimental study of the mechanical contact behavior, under monotonic and cyclic spherical indentation, of an AlCrN film coated onto two distinct hard tool substrates: cemented carbides and tool steel, the following conclusion can be drawn:

1. Circular cracking at the coating is a more appropriate choice than interfacial delamination for defining critical damage under spherical indentation for the coated hardmetal studied here. Such a statement is based on the fact that it not only emerges as the first failure feature observed with increasing load but also leads to the substrate cracking without any intermediate failure at the interface.

2. Coated systems studied here are susceptible to "real" contact fatigue. However, the substrate plays an important role in the failure-related contact response of the coated system. Cemented carbides

represents a harder and stiffer substrate than tool steel, hence, it deforms less and enables the upward coating easier to follow its deformation before the coating experiencing brittle rupture. On the other hand, once crack initiate, tool steel represents a tougher substrate, thus better equipped to arrest or inhibit any penetrating cracks.

Acknowledgements

This work was financially supported by Spanish Ministerios de Ciencía e Innovación (oPtiMa-IPT-2011-1376-020000) and Economía y Competitividad (Grant MAT2012-34602). One of the authors (J.Y.) acknowledges funding received through Erasmus Mundus joint European Doctoral Programme DocMASE. Additionally, SEM and FIB assistance by I.López and T.Trifonov are kindly acknowledged.

References

[1] Mo J L, Zhu M H, Lei B, Leng Y X and Huang N 2007 *Wear* **263** 1423

[2] Mo J L, Zhu M H, Leyland A and Matthews A 2013 *Surf. Coat. Technol.* **215** 170

[3] Mitterer C, Holler F, Reitberger D, Badisch E, Stoiber M, Lugmair C, Nobauer R and Kullmer R. 2003 *Surf. Coat. Technol.* **163-164** 716

[4] Fox V, Jones a, Renevier N and Teer D 2000 *Surf. Coat. Technol.* **125** 347

[5] Podgornik B, Hogmark S and Sandberg O 2004 *Surf. Coat. Technol.***184** 338

[6] Carlsson P and Olsson M 2006 *Surf. Coat. Technol.* **200** 4654

[7] Taylor P and Guiberteau F 1993 *Philos. Mag.* **68** 37

[8] Lawn B R 1998 *J. Am. Ceram. Soc.* **94** 1977

[9] Lawn B R, Deng Y, Miranda P, Pajares A, Chai H and Kim D K 2002 *J. Mater. Res.* **17** 3019

[10] Tarrés E, Ramírez G, Gaillard Y, Jiménez-Piqué E and Llanes L 2009 *Int. J. Refract. Met. Hard Mater.***27** 323

[11] Ramírez G, Mestra A, Casas B, Valls I, Martínez R, Bueno R, Góez A, Mateo A and Llanes L 2012 *Surf. Coat. Technol.* **206** 3069

[12] Góez A, Coureaux D, Ingebrand A, Reig B, Tarrés E, Mestra A, Mateo A, Jiménez-Piqué E and Llanes L 2012 *Int. J. Refract. Met. Hard Mater.* **30** 121

[13] Hogmark S, Jacobson S and Larsson M 2000 *Wear* **246** 20

[14] Bantle R and Matthews A 1995 *Surf. Coat. Technol.* **74-75** 857

[15] Jayaram V, Bhowmick S, Xie Z-H, Math S, Hoffman M and Biswas S K 2006 *Mater. Sci. Eng.* A **423** 8

[16] Cairney J, Tsukano R, Hoffman M and Yang M 2004 *Acta Mater.***52** 3229

[17] Abdul-Baqi A and Van der Giessen E 2002 *Int. J. Solids Struct.* **39** 1427

IOP Conf. Series: Materials Science and Engineering **48** (2013) 012004 doi:10.1088/1757-899X/48/1/012004

Method for the evaluation of stretch blow molding simulations with free blow trials

Johannes Zimmer, Markus Stommel
Chair of Polymer Materials, Saarland University, Saarbrücken, Germany

E-mail: j.zimmer@mx.uni-saarland.de

Abstract. Finite-Element (FE) simulations are a valuable tool to support the analysis and optimization of production processes. In order to achieve realistic simulation results, a consistent simulation set-up followed by an evaluation through experiments is crucial. Stretch Blow Molding (SBM) is a commonly applied forming method to produce thin walled bottles. Polyethylene terephthalate (PET) preforms are biaxially stretched into a closed cavity to form a bottle. In this process the thermo-mechanical material behavior during forming greatly influences the performance of the end product and consequently plays a key role for a reliable process simulation. To ensure a realistic material representation in the simulation model, an adequate material model is calibrated with stress-strain curves from biaxial tests. Thin PET-sheets are stretched under defined temperatures and strain rates. These representative experiments include process simplifications regarding geometry, heating and deformation parameters. Therefore, an evaluation step subsequent to the simulation set-up is inevitable. This paper presents a method for extracting temperature dependent stress-strain-curves from experiments close to the production process which enables the crucial evaluation of a process simulation. In the SBM process, the wall thickness distribution of the bottle refers to the preform deformation over time but does not fully define the thermo-mechanical material behavior. In the presented method, PET-preforms receive thermal treatment with Infrared (IR)-heaters from an SBM-machine and are subsequently inflated into free air (free-blow-trial). An IR-camera is used to obtain the temperature distribution on the preform immediately before blowing. Two high speed cameras are synchronized with a pressure sensor to consequently calculate reliable stress-strain curves at any point on the preform surface. These data is finally compared to results from FE-simulations of the free blow trials.

1. Introduction

Stretch Blow Molding (SBM) is a high speed production process to form plastic bottles and containers. The most prevalent material used in this process is poly(ethylene terephthalate) (PET). In the two stage process, injection molded amorphous PET (aPET)-preforms are first heated (heating stage) in an Infrared (IR) oven above their glass transition temperature (T_g, around 85°C). In a subsequent step the preforms are biaxially stretched (blowing stage) with high strain rates into a cavity of the desired bottle shape. One of the advantages of PET compared to other materials is the strain hardening effect due to molecular orientation at high strain ratios. This hardening results in an increased biaxial strength and dimensional stability and is dependent on the physical parameters: strain rate, temperature and deformation mode. These parameters are affected by the process conditions. The heating time and the infrared lamps power create the desired axial and radial temperature profile of the preform and the fluid pressure and the stretch speed determine the strain rate and the strain mode, respectively. Hence, the resulting bottle performance, e.g. represented by its top load behavior, is highly dependent on these process conditions. Although the process itself is well known, the increasing demands for lightweight bottles and a better bottle performance pose a permanent challenge.

Nowadays FE-simulations can replace extensive testing to find process parameters leading to a better bottle performance. The thermo-mechanical characterization of PET plays a key role for the simulation of a stretch blow molding process. A lot of research has been done so far to capture the complex material behavior during SBM or related processes like thermoforming and extrusion blow molding. In early approaches the material is regarded as purely hyperelastic [1-5], since PET and other thermoplastics show a rubberlike behavior above T_g. However, these approaches cannot cover the strong rate and temperature dependence of PET. Therefore, more recently developed models are based on the attempt of incorporating these effects by using visco-plastic [6-8], visco-elastic [9-11] or visco-hyperelastic [12, 13] approaches. For example, the approach from Buckley et al. [14, 15] is a physically based visco-hyperelastic material model. Adams et al. developed it further [16] to achieve a better prediction of the temperature dependent material behavior at high strains. Several workgroups successfully applied this model for predicting the wall thickness distribution of a stretch blow molded bottle. In comparison to the above mentioned models, a good compromise between numerical stability, accuracy in the wall thickness prediction and calculation time is obtained.

A prediction of the real stress-strain behavior depends not only on the choice of the constitutive equations. Furthermore the determination of the specific constants (material parameters) is important. In practice, these parameters are identified by fitting a constitutive equation to experimental data. Hereby, it is essential that the data are created in conditions close to the subsequent application of the material model. During SBM stress-strain data can be hardly obtained. The physical parameters temperature, strain rate and deformation mode are determined by the process conditions but cannot be set up to defined values. The preforms' temperature profile results from the infrared heating. During the blowing process the strain rate varies in time and depends on the heating and the load conditions. Moreover the preforms are enclosed by a cavity during the deformation which complicates the measurement. Therefore representative tests are used to determine the material parameters. Through a defined variation of temperature, strain rate and deformation mode between the tests, the thermo-mechanical material behavior is characterized. For SBM, a common approach is to perform biaxial stretching experiments [17-19]. Thin PET sheets are heated up to a certain temperature and afterwards deformed with a constant stretch speed in one or two directions. A machine that serves this purpose is the biaxial stretcher at Queens University in Belfast [18]. During stretching the strain is calculated by the displacement of the clamps. The nominal stress is determined from the reaction force and the initial measures of the plates. However, these experiments include process simplifications regarding geometry, heating and deformation parameters. The production of PET sheets could imply different internal stresses compared to the preforms due to different shear loadings during injection molding. Moreover the experiments do not cover the full field of application in terms of temperature and strain ratios in SBM, because it is difficult to stretch thin PET sheets at temperatures below T_g up to high stretch ratios without material failure.

An indispensable step to obtain a reliable material characterization is therefore a validation of the so created material parameters through experiments corresponding to the production process. Menary et al. validated a material model through free blow trials [20]. They recorded the geometry of preforms during deformation in free air. These trials were afterwards FE-simulated with a model that was calibrated by biaxial stretching experiments. Simulation and experiments showed a comparable preform shape evolution at equal times.

To further evaluate the quality of the model and the material parameters, it is desirable to extract stress-strain curves from SBM or related experiments like free blow trials. In contrast to stretching PET sheets, contact measurements could affect the free deformation of the PET, which requires optical measurements. Related investigations have been done for different materials than PET. Charalambides et al. used the bubble inflation technique to determine the stress-strain behavior of dough [21, 22]. In this method a regular pattern is assigned on a flat membrane which is clamped and sealed at the perimeter. Air inflates the membrane from one side and the resulting deformation is recorded by a camera system, which enables the extraction of the strain in the pole of the membrane. The membrane's current thickness is here approximated by an analytical function. In combination with the pressure measured during the deformation and under the assumption of incompressibility this enables the calculation of the stress in the pole. Reuge et al. used the bubble inflation method to investigate the biaxial behavior of rubber [23]. The inflation process was captured with a frame rate of 24/s. Additionally, the current thickness during deformation was measured using a magnetic probe. It was found that, assuming an analytical function for the thickness, does not yield acceptable results.

In the following procedure the optical strain measurement is adopted to the SBM process and extended. A three-dimensional camera system and a therewith synchronized pressure sensor allow the determination of the stress-strain information at any point on the preform surface. Additionally, an IR-camera allows recording the preform surface temperature at the same point. From this data, a thickness temperature profile is calculated and linked with the stress-strain information to create a reliable characterization of the PET's thermo-mechanical behavior in practice. To use this data as an evaluation an FE-simulation of the free blow trials is set-up. Hereby, the material model from Buckley et al. [14, 15] and Adams et al. [16] is used. To obtain the material parameters, the model is calibrated with data from stretching experiments of PET sheets. A comparison of resulting stress-strain-curves and the deformation at different time steps from free blow simulations and free blow trials evaluate the applicability of the material model to the SBM process as well as the free blow simulation model.

2. Experiments

2.1. Stretching of PET sheets

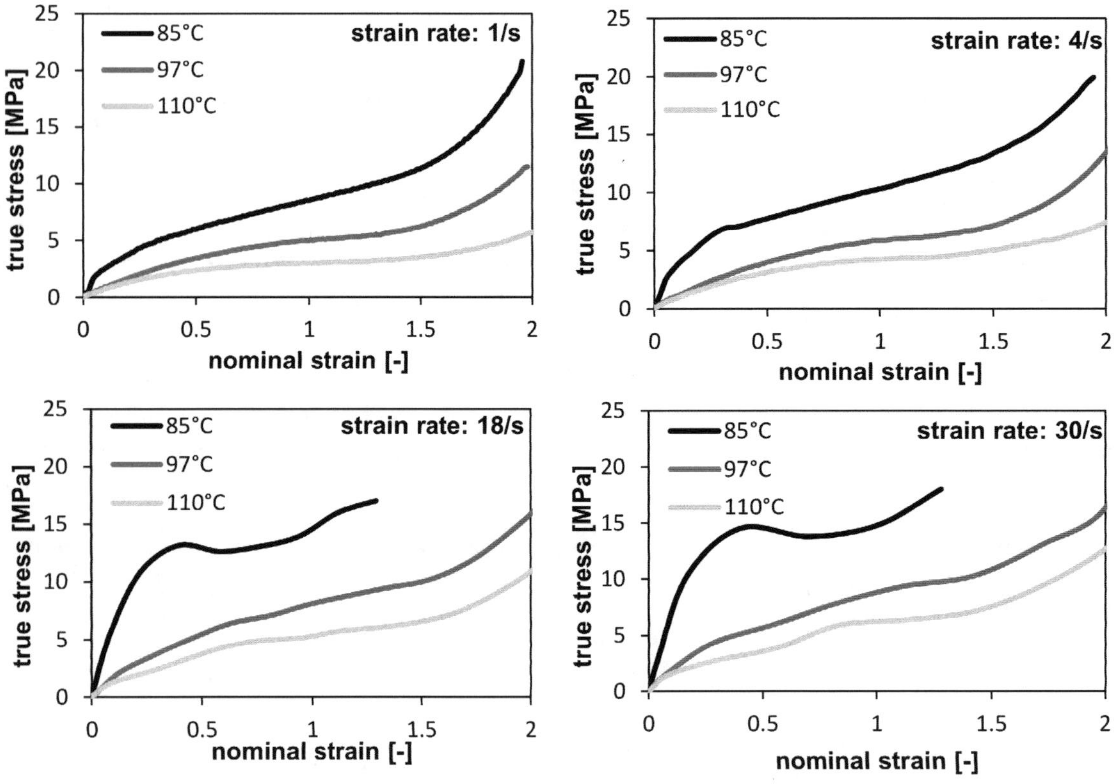

Figure 1. Stress-strain results from equibiaxial stretching of PET sheets.

Stretching experiments of injection molded aPET sheets are performed at the biaxial stretcher from Queen's university. The sheets have lateral extensions of 76mm x 76mm and a thickness of 1.2mm. A Dow LIGHTER™ C88 PET resin is used. The Experiments are conducted at temperatures from 85°C to 110°C and nominal strain rates range from 100% per second (1/s) to 3000% per second (30/s). Tested deformation modes are equibiaxial and sequential biaxial. Representative stress-strain curves

IOP Conf. Series: Materials Science and Engineering **48** (2013) 012004 doi:10.1088/1757-899X/48/1/012004

are shown in Figure 1. The true stress was thereby calculated under the assumption of incompressibility by the clamping force and the stretch ratio in thickness direction.

2.2. Free blow trials

A schematic of the infrared oven and the free blow test rig is shown in Figure 2. The PET resin used for preform injection molding is the same as for the calibration experiments in 2.1. The aPET preforms are heated in an infrared oven equipped with six infrared lamps to create temperature profiles comparable to the real production process. For the free blow trials, IR-lamps 1-3 are switched off so that only the part of the preform heated by IR-lamps 4-6 deforms, which is mainly the preform dome. As a consequence, the deformation is more stable and the cameras can have a closer view on the area of interest due to a smaller overall deformation. The heated preform is transferred to the free blow test rig where it is clamped and exposed to a pneumatic pressure to blow it into free air. Two high speed cameras record the preform deformation during blowing with a frame rate of 2000/s. The corresponding pressure is obtained by a dynamic sensor. The short deformation times require real time control and data acquisition which is done by a Programmable Automation Controller (PAC) device. A FPGA chip is programmed using the software package LabVIEW to obtain high speed control. The PAC synchronizes the pressure sensor with the cameras to receive for each camera image the corresponding pressure value.

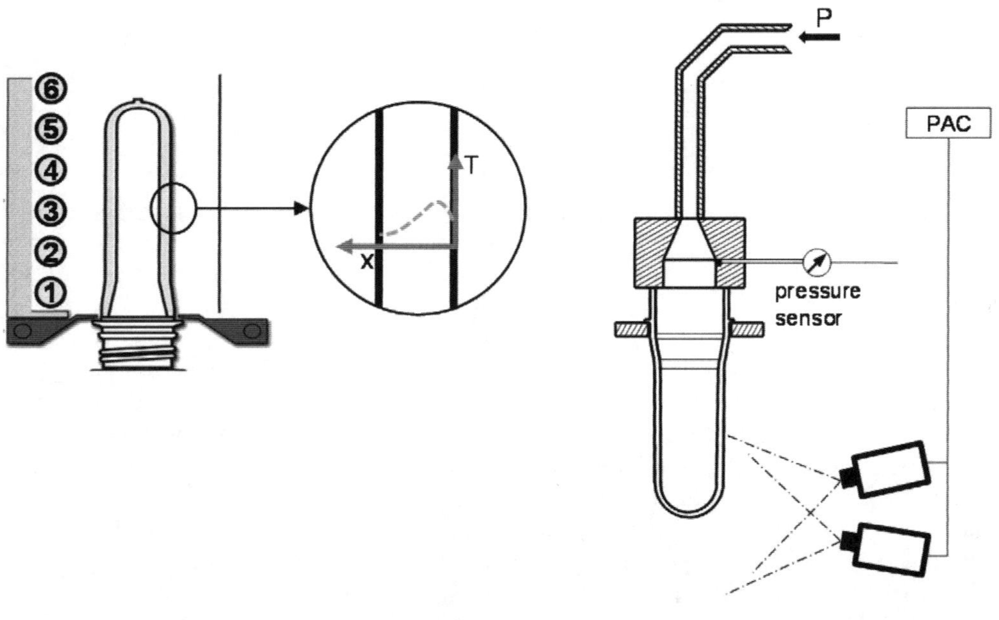

Figure 2. Experimental set up, left: IR-oven, right: free blow test rig.

The images recorded during blowing are evaluated with digital image correlation. The advantage here compared to previous works [21, 23] is the three dimensional measurement. Every point on the preform surface is observed by two cameras from different directions. The positions of the two cameras relatively to each other are known from a calibration, which makes it possible to calculate the absolute three dimensional coordinates of the preform surface in space at every point in time. This procedure works only, when the object surface shows enough structure to allow the algorithms to correlate identical points from both cameras. Hence, the preforms are prepared with a random speckle pattern which proved not to influence the deformation behavior. The grey values are tracked in small neighborhoods and through a correlation of the images, the three dimensional strains are calculated. The pattern correlation allows the determination of the principle in-plane-strains and the radius of curvature in the same directions. These data are calculated by the software Vic3d at almost every point on the preform surface detected by the cameras. A more detailed description of the mathematical calculation procedure can be found in [24]. Figure 3 shows the pressure progression during free blowing with a heating time of 24s. At three times (0s, 0.1s and 0.2s) the corresponding preform shape is displayed.

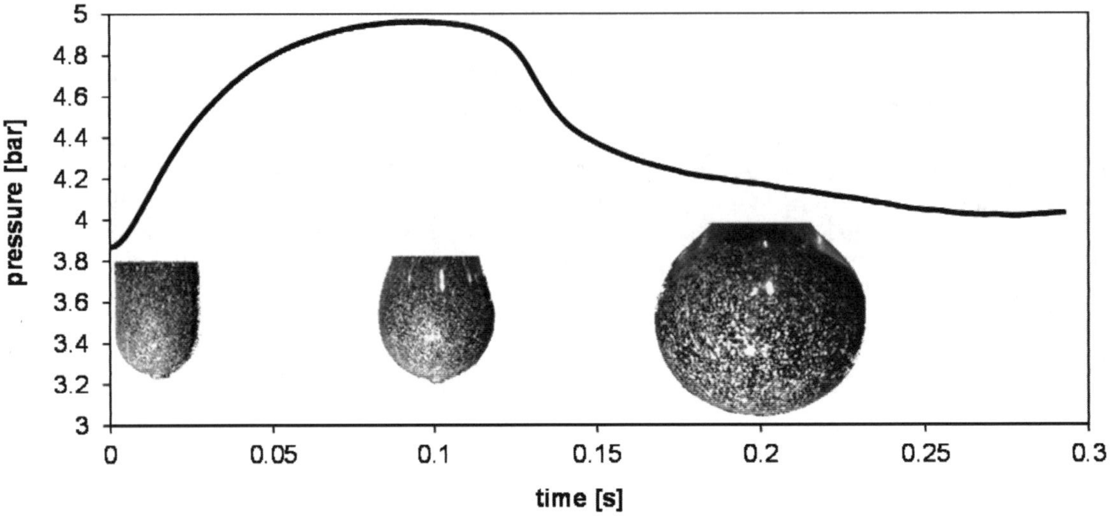

Figure 3. Pressure during free blowing and corresponding preform shape.

The heating set-up creates a temperature profile constant in the circumferential direction but varying in the axial- and the thickness-direction (see Figure 2, left). Furthermore, the temperature varies over time because of cooling effects. To evaluate the material model, a temperature value for the point of stress-strain calculation is required. This value has to take into account the thickness variation and the temperature decrease. In order to include these effects, measurements with an infrared camera are conducted. The camera records the surface temperature evolution over 20 seconds with a frame rate of 5/s for the complete body. The measurement starts directly after the preform exits the IR oven. From this data the temperature evolution in time at any point on the outer surface can be determined (see Figure 4, left). The time course of temperature is fit to an analytical function, which serves as input data for an algorithm to calculate the temperature distribution in the thickness direction at this point (Figure 4, right). A detailed principle of the procedure is shown in [25]. To achieve reliable results, the algorithm is proofed by an additional measurement at the inner surface of the preform. A correlation with the stress-strain data is done by calculating the mean temperature over thickness after the specific transfer time (about 2 seconds). The methods explained above finally lead to the determination of stress-strain data for different temperatures. This thermo-mechanical data can be created on the one side by measuring different points on the preform-surface. Alternatively, the influence of process parameters can be investigated by calculating the stress-strain-information in multiple trials at the same position on the surface after different heating configurations. This is shown in Figure 5, where the heating time is varied (20-28s). The advantage compared to the stretching experiments of PET sheets is, that from one trial multiple stress-strain curves for different temperatures (different points on the surface) can be extracted. Furthermore, it is difficult to deform thin PET sheets up to high stretch ratios at low temperatures ($<T_g$). Hence, the material model may not realistically describe the PET behavior in this area. On the contrary, during blowing in free air, points on the preform surface with temperatures below T_g can be measured up to high stretch ratios due to the decreasing temperature gradient from the preform dome towards the neck.

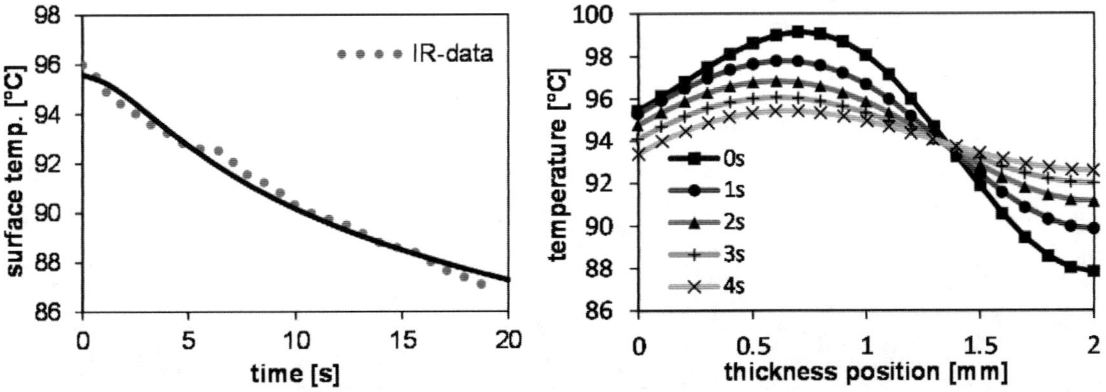

Figure 4. Temperature determination, left: IR-camera data and fit at point on outside surface, right: temperature profile over thickness at the same point, (0: outside surface).

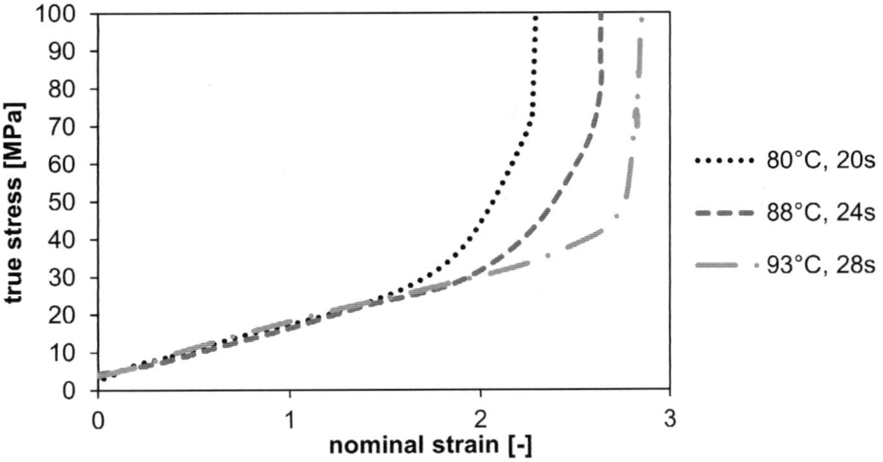

Figure 5. Stress-strain curves extracted from free-blow-trials.

3. FE-Simulation

3.1. material model validation

In this study the so called 'glass-rubber constitutive model' developed by Buckley et al. [14] is implemented into an FE-package. The material parameters are identified according to Adams [16]. An explicit FE-solver is used because it is more stable than an implicit solver for the large and fast deformations arising during SBM. The constitutive equations are implemented in the user material subroutine VUMAT of the FE-package ABAQUS/Explicit. The model is validated via the VUMAT in ABAUS/Explicit. A detailed description of the implementation procedure is given in [24]. Stress-strain data extracted from the stretching experiments and the VUMAT are compared in Figure 6.

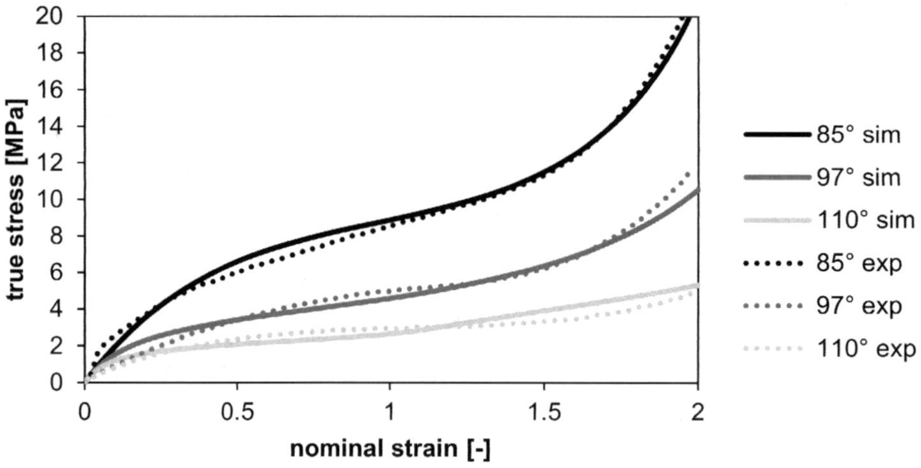

Figure 6. Results from equibiaxial experiments and simulations, strain rate: 1/s.

3.2. Process model evaluation

The stress-strain curves from Figure 5 show the thermo-mechanical behavior of PET under process related conditions. These data cannot directly be compared to those from the calibration (Figure 6) because, as in SBM, the strain rate is not constant during the free blow trials. The material parameter determination in [15, 16] operates with stress-strain data at defined strain rates. Therefore, the material model, calibrated with stretching of PET-sheets, is implemented in an FE-simulation of the free blow trials. In this way the material model and the free blow simulation can be evaluated. The heating configuration of the free blow trials leads to a deformation only of the lower preform part. A large part of the preform stays below the glass transition temperature. Consequently only the part where a deformation can occur (40mm from the preform pole towards the neck) is modeled with the calculated temperature profile. The "fluid cavity" approach in ABAQUS/Explicit is used to simulate the air flow into the preform. The measured pressure evolution during time (see Figure 3) is adjusted via a mass flow rate in the simulation.

Figure 7 compares the preform deformation derived from experiment and FE-simulation at equal times. It can be seen that the preform shape is almost identical in both cases. In Figure 8, resulting stress strain data at an exemplary point on the preform surface at 88°C is extracted from experiment and simulation. The curves match well in a large area of strain. However, the initial material stiffness is slightly underestimated by the simulation. This is probably due to the soft material behavior in the calibration trials for small strains, (see Figure 1). Further deviation seems to occur in the area of very large strains, where no experimental data from the stretching experiments exists. However for SBM this is the area of interest where hardening effects appear resulting in the final product performance in terms of material behavior. As a consequence, the material behavior at high strains investigated with free blow trials needs to be implemented into the FE-simulation.

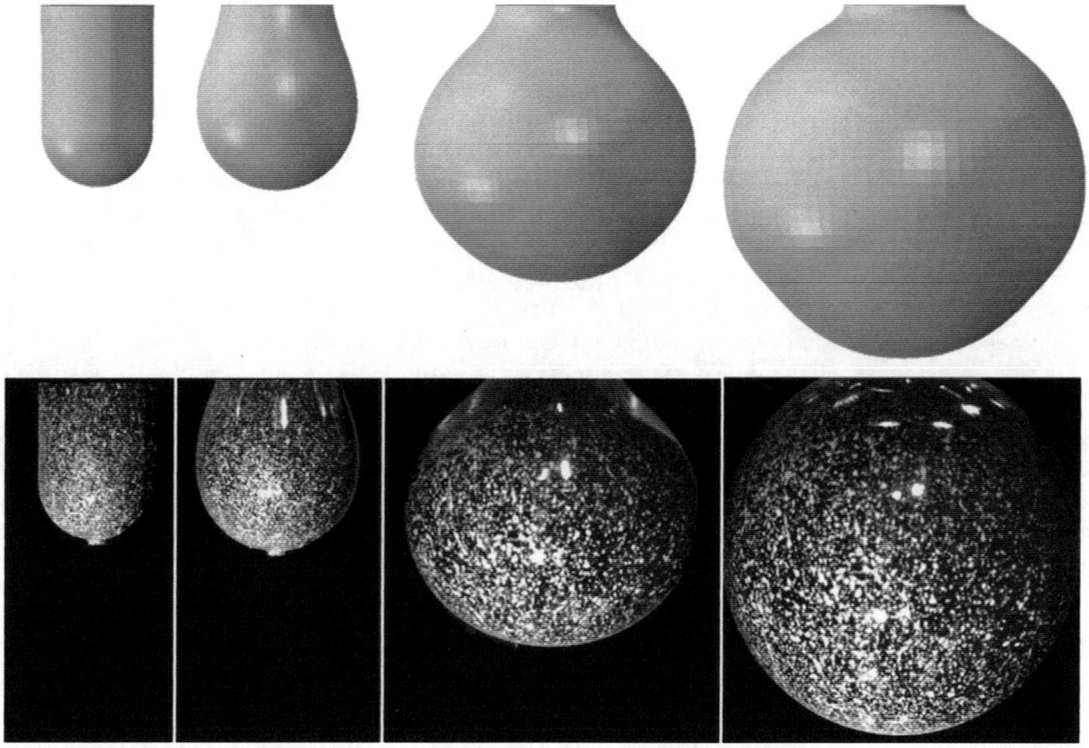

Figure 7 comparison of preform shape from simulation (up) and experiments (down).

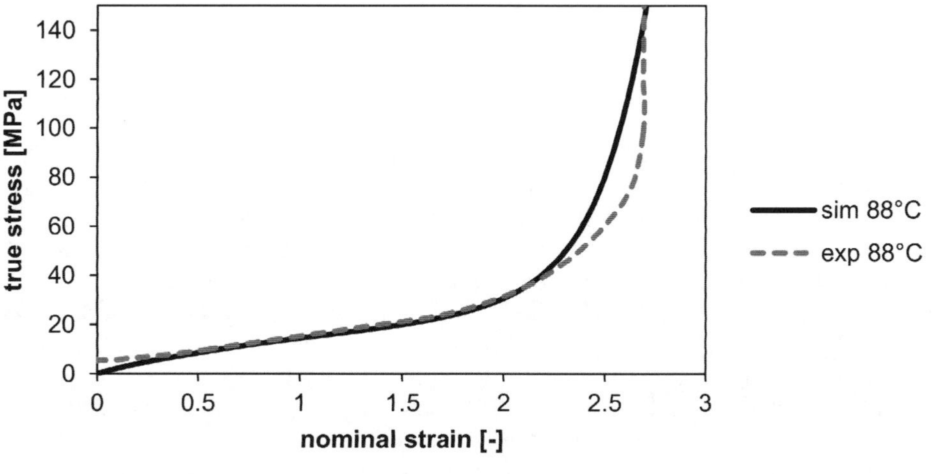

Figure 8: free blow-trials, comparison of measured and simulated results.

4. Conclusions

In this paper a method is presented to evaluate a material model and an FE-simulation for stretch blow molding. As it is common practice, the model is calibrated with stress-strain data from stretching thin PET sheets at different temperatures. Since these experiments imply simplifications and do not cover the full field of application, the capabilities and limits of the model are investigated. Therefore stress-strain data at different temperatures are extracted from free blow trials, which are closer to SBM. This is done with an optical high speed measurement system, which enables the extraction of multiple stress-strain curves at different temperatures from a single free-blow-trial. The resulting curves are compared to FE-simulations of the free-blow trials. Hereby the material model, calibrated with data from stretching PET sheets, is implemented.

Experiments and Simulations match well in terms of preform deformation and stress-strain data (see Figure 7 and Figure 8). This indicates that the material model is well coded and is able to cover the thermo-mechanical behavior of PET. Therefore, the stretching experiments can be used to calibrate the material model. Furthermore the FE-model of the free blow trials can be extended to realistically simulate the stretch blow molding process. However, the stress strain curves from experiment and simulation deviate at the beginning and for very large strains. The area of low strains should be carefully investigated in a further study to identify the cause of this difference. The deviation for very high strains (>200%) indicates that the material model is not able to predict the PET behavior in this area, because no experimental data for calibration was provided in this section.

Consequently, in a following work the stress strain information from free-blow trials should directly be used to calibrate the material model to cover the behavior of PET in SBM for higher strain ratios and low temperatures. For this purpose the changing strain rate during blowing has to be taken into account. To further optimize the simulation method, the temperature profile over the thickness has to be integrated in the FE-simulation. In this study only a mean thickness temperature is calculated from the profile. Furthermore, a stretch rod in the free blow trials would contribute to even better represent the real production process.

5. Acknowledgments

This work was conducted within an industrial project with Nestlé Waters, whose funding is gratefully acknowledged. The testing and discussions with Guillaume Chauvin and Damien Kannengiesser from Nestlé Waters and Marc Max Schöneich from University Saarland are deeply appreciated. Further thanks to Gary Menary from Queen's University in Belfast for conducting the stretching experiments.

6. References

[1] Kouba K, Bartos O, and Vlachopoulos J 1992 *Computer-Simulation of Thermoforming in Complex Shapes* Polymer Engineering and Science 32 pp 699-704

[2] Marckmann G, Verron E, and Peseux B 2001 *Finite Element Analysis of Blow Molding and Thermoforming Using a Dynamic Explicit Procedure* Polymer Engineering and Science 41 pp 426-439

[3] Carlone P, and Palazzo G S 2006 *Finite Element Analysis of the Thermoforming Manufacturing Process Using the Hyperelastic Mooney-Rivlin Model* Computational Science and Its Applications - Iccsa 2006, Pt 1 3980 pp 794-803

[4] Nied H F, Taylor C A, and Delorenzi H G 1990 *Three-Dimensional Finite Element Simulation of Thermoforming* Polymer Engineering and Science 30 pp 1314-1322

[5] Briatico-Vangosa F, Rink M, D'Oria F, and Verzelli A 2000 *Deformational Behavior of Polyolefins at High Temperature and Strain Rate: Experimental Analysis and Constitutive Laws* Polymer Engineering and Science 40 pp 1553-1563

[6] Cosson B, Chevalier L, and Regnier G 2012 *Simulation of the Stretch Blow Moulding Process: From the Modelling of the Microstructure Evolution to the End-Use Elastic Properties of Polyethylene Terephthalate Bottles* International Journal of Material Forming 5 pp 39-53

[7] Wang S, Makinouchi A, and Nakagawa T 1998 *Three-Dimensional Viscoplastic Fem Simulation of a Stretch Blow Molding Process* Advances in Polymer Technology 17 pp 189-202

[8] Bordival M, Schmidt F M, Le Maoult Y, and Velay V, "Simulation of the Two Stages Stretch-Blow Molding Process: Infrared Heating and Blowing Modeling," *AIP Conference Proceedings*, American Institute of Physics, 2007, pp. 519-524

[9] Debbaut B, Homerin O, and Jivraj N 1999 *A Comparison between Experiments and Predictions for the Blow Molding of an Industrial Part* Polymer Engineering & Science 39 pp 1812-1822

[10] Laroche D, Kabanemi K K, Pecora L, and Diraddo R W 1999 *Integrated Numerical Modeling of the Blow Molding Process* Polymer Engineering and Science 39 pp 1223-1233

IOP Conf. Series: Materials Science and Engineering **48** (2013) 012004 doi:10.1088/1757-899X/48/1/012004

[11] Tanifuji S I, Kikuchi T, Takimoto J I, and Koyama K 2000 *Overall Numerical Simulation of Extrusion Blow Molding Process* Polymer Engineering and Science 40 pp 1878-1893

[12] Dupaix R B, and Boyce M C 2007 *Constitutive Modeling of the Finite Strain Behavior of Amorphous Polymers in and above the Glass Transition* Mechanics of Materials 39 pp 39-52

[13] Pham X T, Thibault F, and Lim L T 2004 *Modeling and Simulation of Stretch Blow Molding of Polyethylene Terephthalate* Polymer Engineering and Science 44 pp 1460-1472

[14] Buckley C P, and Jones D C 1995 *Glass-Rubber Constitutive Model for Amorphous Polymers near the Glass Transition* Polymer 36 pp 3301-3312

[15] Buckley C P, Jones D C, and Jones D P 1996 *Hot-Drawing of Poly(Ethylene Terephthalate) under Biaxial Stress: Application of a Three-Dimensional Glass—Rubber Constitutive Model* Polymer 37 pp 2403-2414

[16] Adams A M, Buckley C P, and Jones D P 2000 *Biaxial Hot Drawing of Poly(Ethylene Terephthalate): Measurements and Modelling of Strain-Stiffening* Polymer 41 pp 771-786

[17] Marco Y, Chevalier L, Regnier G, and Poitou A 2002 *Induced Crystallization and Orientation of Poly(Ethylene Terephthalate) During Uniaxial and Biaxial Elongation* Macromolecular Symposia 185 pp 15-34

[18] Martin P J, Tan C W, Tshai K Y, McCool R, Menary G, Armstrong C G, and Harkin-Jones E M A 2005 *Biaxial Characterisation of Materials for Thermoforming and Blow Moulding* Plastics Rubber and Composites 34 pp 276-282

[19] Menary G H, Tan C W, Harkin-Jones E M A, Armstrong C G, and Martin P J 2012 *Biaxial Deformation and Experimental Study of Pet at Conditions Applicable to Stretch Blow Molding* Polymer Engineering and Science 52 pp 671-688

[20] Menary G H, Tan C W, Armstrong C G, Salomeia Y, Picard M, Billon N, and Harkin-Jones E M A 2010 *Validating Injection Stretch-Blow Molding Simulation through Free Blow Trials* Polymer Engineering and Science 50 pp 1047-1057

[21] Charalambides M N, Wanigasooriya L, and Williams J G 2002 *Biaxial Deformation of Dough Using the Bubble Inflation Technique. Ii. Numerical Modelling* Rheologica Acta 41 pp 541-548

[22] Charalambides M N, Wanigasooriya L, Williams J G, and Chakrabarti S 2002 *Biaxial Deformation of Dough Using the Bubble Inflation Technique. I. Experimental* Rheologica Acta 41 pp 532-540

[23] Reuge N, Schmidt F M, Le Maoult Y, Rachik M, and Abbe F 2001 *Elastomer Biaxial Characterization Using Bubble Inflation Technique. I: Experimental Investigations* Polymer Engineering and Science 41 pp 522-531

[24] Zimmer J, Detrois C, and Stommel M 2013 *Evaluation Method for Stretch Blow Moulding Simulations with Process-Oriented Experiments* Key Engineering Materials 554-557 pp 1658-1668

[25] Shelby M D, Horton K D, and Hall H P, *Method and Device for Predicting Temperature Profiles Throughout the Thickness of a Polymer Preform*, US2004/0024560A1, 2004.

IOP Conf. Series: Materials Science and Engineering **48** (2013) 012005 doi:10.1088/1757-899X/48/1/012005

Tie molecules, morphology and confinement effects in semi – crystalline poly(ethylene naphthalene-2,6-dicarboxylate) (PEN)

E Nikaj[1], I Royaud[2,3] and L David[1]

[1]Université de Lyon, Université Lyon 1, UMR CNRS 5223 Ingénierie des Matériaux Polymères (IMP), Laboratoire IMP@LYON1, Bât. PolytechLyon, 15, bd. Latarjet F-69622 Villeurbanne Cx, France
[2]Université de Lorraine, Institut Jean Lamour, UMR CNRS 7198, Département SI2M Parc de Saurupt- CS 50840 ; 54 011 Nancy Cx, France
[3]EEIGM, 6 rue Bastien-Lepage, BP 10630, 54 010 Nancy Cx, France

*Corresponding author : I. Royaud, Tel : + 33 383584143; Fax: + 33 383584056
e-mail: isabelle.royaud@univ-lorraine.fr

Abstract. The glass transition dynamics and confinement effects in poly(ethylene naphthalene-2,6-dicarboxylate) (PEN) were studied by broadband dielectric relaxation spectroscopy (DRS) in relation with semi-crystalline morphology. PEN samples were obtained by cold crystallization at different crystallization temperatures (T_c ranging from 165 to 240°C) and crystallization times (t_c from 30min to 24h). Differential scanning calorimetry and X-ray diffraction showed that the crystallinity ratio (X_c) increases when T_c and t_c increase. The glass transition relaxation is shifted to higher temperatures as t_c increases but this confinement effect decreased with increasing T_c. The origin of this anomalous dynamics can be related to (*i*) the crystalline lamellar stack morphology revealed by small angle X-ray scattering (SAXS) and (*ii*) chain scission occurring during annealing at the crystallization temperature. As a result, the density of tie-chain molecules (chain portions bridging crystallites) can be considered as the key factor for the understanding of confinement effects in semicrystalline polymers. Accordingly the confinement effects on the glass transition dynamics can be optimally rationalized as a function of a characteristic length intermediate between the interlamellae thickness l_a and that of the theory of Brown and Huang $2l_c+l_a$ where l_c is the crystallite thickness.

1. Introduction

The molecular mobility in semi-crystalline polymers has been studied by several authors using a large number of systems and spectroscopic methods [1-18]. However, there are still few works analyzing simultaneously in details their molecular and nano structure in relation with their dynamics [9-12]. As a general trend, these studies have commonly revealed an increase in glass transition temperature T_g when the crystallinity ratio is sufficiently high. Obviously, this trend is attributed to the slowing down of the molecular mobility in the amorphous phase that is constrained by the presence of the crystalline phase. The confinement effect is due to the participation of the chains within the crystals and can also result in the formation of a rigid amorphous fraction (RAF)[6,7,19-21]. Cheng and Wunderlich[20] previously showed that the glass transition of semi-crystalline PEN increases and an increasing RAF is developed with decreasing crystallization temperatures. Also, in specific crystallization conditions ("constrained" conditions [18]) a different behavior can be observed as a result of a lowering of the density of the amorphous phase. The dynamics of the amorphous phase thus appear to result both from the confinement effects and from the amorphous structural state inherited from crystallization conditions. In addition, confinement was also envisaged to influence the lamellar stacks structuring itself in the crystallization process[22,23].

In this context, a general understanding of the relations between thermo-mechanical history, structure and dynamics is still lacking to completely understand the dynamic behavior of these complex systems. In this paper, we have chosen to study the molecular dynamics of Poly(ethylene naphthalene-2,6-dicarboxylate) (PEN) because of the large range of crystallization temperatures that can be assessed to induce crystallization from the amorphous films. PEN is a thermoplastic polyester developed in the 1970s. It combines thermal stability, degradation resistance, good barrier properties and excellent dielectric and mechanical properties[5]. The melting point temperature of PEN ($T_m \approx$ 267°C, measured at a heating rate of 10°C/min) obtained by differential scanning calorimetry (DSC) is about 10°C higher than in the case of PET, while the glass transition temperature ($T_g \approx 120$°C) is about 40°C higher than that for PET [24]. The enhanced properties of PEN compared to PET are due to a higher stiffness of the macromolecular chain resulting from the presence of a naphthalene structure (see Scheme 1) [25]. Thus, PEN finds a large variety of applications as tire cords for cars[26], as electromagnetic tapes (audio, video and computer tapes) [27, 28] and in packaging or capacitors industry.

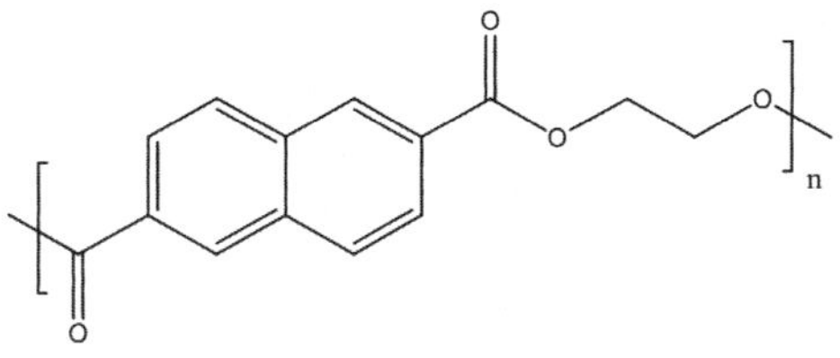

Scheme 1: Chemical structure of poly(ethylene-2,6-naphthalene dicarboxylate)

Several studies on the molecular mobility or the morphology of PEN have already been published[2,11,12]. However, few of them successfully bridge the gap between dynamics and morphology at the nanoscale. The aim of this work was to illustrate the links between thermal history, semi-crystalline morphology and confinement of the amorphous phase by the crystalline lamellae. The α-relaxation dynamics is the most affected by crystallinity. Therefore, this paper will deal mainly with the study of α-process in PEN as a function of crystallization conditions. Previous works have shown that α-relaxation of PEN classically shifts to higher temperatures as the result of the confinement of the amorphous phase by the crystalline phase[4-7]. In addition to the development of the semi-crystalline morphology, a structuration of the amorphous phase is likely to occur with the formation of a rigid amorphous phase, *i.e.* an amorphous contribution that do not contribute to the specific heat increment at the glass transition[7]. Such confinement effects in polymer materials generally lead to a change in the dynamics in the amorphous phase, but slower, faster or unchanged confined dynamics have previously been described[29] as a function of the interactions between the rigid confining phase and mobile polymeric phase. This phenomenon is generally probed by the evolution of the calorimetric glass transition or the frequency/temperature location of the α-relaxation process (dielectric, mechanical…etc). A derived issue is the relation between the confinement effects and the details of molecular mobility in terms of intermolecular cooperativity[30,31]. In this work, the confinement effects are shown to be mainly governed by the cold crystallization conditions at different temperatures and during different crystallization times.

2. Experimental details

2.1. Materials

Amorphous PEN films with a thickness of 109 μm were obtained from Du Pont de Nemours with a weight average molecular weight of about 30100 ± 8000 g/mol. Semicrystalline PEN samples were crystallized in situ in the dielectric cell by heating from room to the crystallization temperatures (T_c = 165, 180, 220 and 245°C) at a rate close to 10 K/min and subsequent annealing during different crystallization times (t_c = 0.5, 2 and 24h). In a second set of experiments, the PEN films were systematically crystallized in an oven at 165, 170, 180, 190, 200, 210, 220, 230, 240 and 245°C during 30 minutes, in order to better follow the influence of the crystallization temperature on the morphology and the resulting molecular mobility. These samples were dried below T_g in the dielectric cell for 2 h at 110°C before the DRS measurements.

2.2. Methods

Broad band dielectric relaxation spectroscopy (BDS) was carried out on the films with circular electrodes having a diameter of 20 mm in parallel plate geometry. A Novocontrol Concept 40 α-analyzer, interfaced to the sample by a Broadband Dielectric Convertor (BDC, Novocontrol), was used to obtain the complex dielectric function:

$$\varepsilon^*(\omega) = \varepsilon'(\omega) - j\varepsilon''(\omega) \quad (1)$$

where ω is the angular frequency, ε' is the real part, ε'' the imaginary or loss part of the complex permittivity. The measurements were performed in a frequency range from 10^{-1} to 10^6 Hz and the temperature of the sample varied from – 140 to 200°C in isothermal steps every 3 K and was controlled under a nitrogen flow with a stability of $\Delta T = 0.1$ K (Novocontrol Quatro System Controller BDS 1330). The data were analyzed by using the Havriliak-Negami (HN) function to fit them, using the Novocontrol software WinFit. The maximum dielectric loss (ε'') was then determined from the fits of each α-relaxation process to yield a macroscopic relaxation time according to:

$$\tau = 1/2\pi f_{max} \quad (2)$$

where f_{max} is the frequency at maximum dielectric loss.

DSC measurements were carried out using a 2920 TA Instruments previously calibrated with indium (with a heating rate of 10 K/min). Sample weights were around 5 to 8 mg. The samples were all sealed in aluminum pans and the melting enthalpies were calculated from the areas of the melting peaks using the linear baselines from 160°C to 290°C, as shown in Figure 1. In order to calculate the crystallinity ratio from calorimetry measurements, the melting heat of the fully crystalline state ΔH_f^∞ was taken equal to 190 J/g[32]. The values of the glass transition temperatures at the midpoint (T_g) were calculated by using a linear estimation of the baseline before (between 90 and 110°C) and after (between 140 and 160°C) the glass transition.

The semi-crystalline nanostructure of the PEN films was studied by SAXS performed at ESRF (Grenoble France) on BM2-D2AM beam-line. The films were placed perpendicularly to the incident beam. The scattered intensity was corrected from empty cell. Although sophisticated approaches are possible for the study of the lamellar stack morphology[23], in this work the semicrystalline nanostructure was simply characterized by the scattering vector (q_{max}) of the maximum of the Lorentz-corrected intensity $I(q).q^2$ in order to deduce the long period, L_p from the Bragg equation $L_p=2\pi/q_{max}$, where $q=4\pi\sin\theta/\lambda$ and λ is the incident wavelength. The crystalline lamellae thickness (ℓ_c) and the amorphous thickness (the intercrystalline thickness ℓ_a) were deduced from the mass crystalline ratio X_c deduced from DSC measurements:

$$\ell_c = L_p \cdot X_c /((\rho_c/ \rho_a)(1- X_c)+X_c) \quad (3)$$
$$\ell_a = L_p\,(1-X_c) /((\rho_a/\rho_c)\,X_c+1-X_c) \quad (4)$$

Where ρ_c=1.407g/cm^2 and ρ_a=1.325g/cm^3 are the mass densities of the (α) crystalline and amorphous phases respectively[33].

3. Results

3.1. Calorimetry analysis and crystallinity characterization

The thermograms of selected samples crystallized during 24 hours at different crystallization temperatures are reported in Figure 1 in the glass transition range. The large differences observed between the thermograms of samples crystallized at 165°C and 245°C are indicative of different semicrystalline morphologies, associated with different crystalline lamellae thicknesses and/or different crystalline-amorphous interfacial energies and/or different crystalline phases.

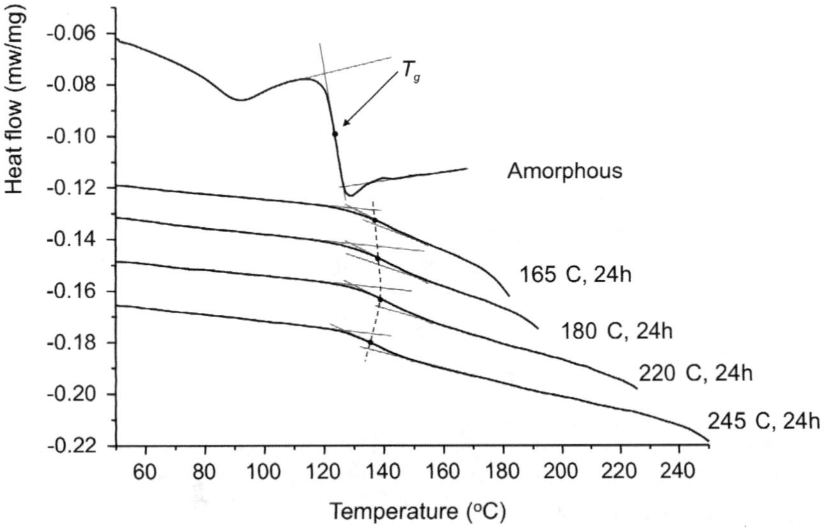

Figure 1: DSC thermograms at 10K/min (0.166 K/s) in the glass transition range for the amorphous and the samples crystallized during 24 hours at 165°C, 180°C, 220°C and 245°C.

A decrease of the glass transition temperatures can be observed after crystallization at 245°C, in Figure 1 in agreement with a previous study[20]. Obviously, these changes in the glass transition dynamics should be related to the differences in the semicrystalline morphology and crystallinity ratios induced by different crystallization conditions. Nevertheless, the glass transition temperatures are difficult to obtain precisely from the thermograms, in particular due to the presence of the secondary endothermic peaks and the low values of the heat capacity increments ΔC_p associated to the

glass transition in semicrystalline systems. More detailed results can be obtained with more sensitive techniques such as BDS.
The effect of crystallinity ratio X_c obtained by DSC can first be discussed in Figure 2, displaying the values of X_c as a function of T_c. Clearly, the decrease in the glass transition temperatures with crystallization temperatures shown in Figure 1 cannot be interpreted as the direct result of the increase of the crystallinity ratio with T_c as observed in Figure 2. The details of the morphology and macromolecular structure must be further analyzed in order to interpret the changes in the dynamics.

Figure 2 (open squares) displays a systematic evolution of crystallinity ratio with crystallization temperature at a constant crystallization time of 30 min. Two crystallization regimes can be distinguished in this figure: at crystallization temperatures below 210°C, the increase of X_c with *Tc* is limited in the range 0.27-0.29. A stronger increase is observed for the samples crystallized above 210°C, with a limiting crystalline ratio just above 0.34.

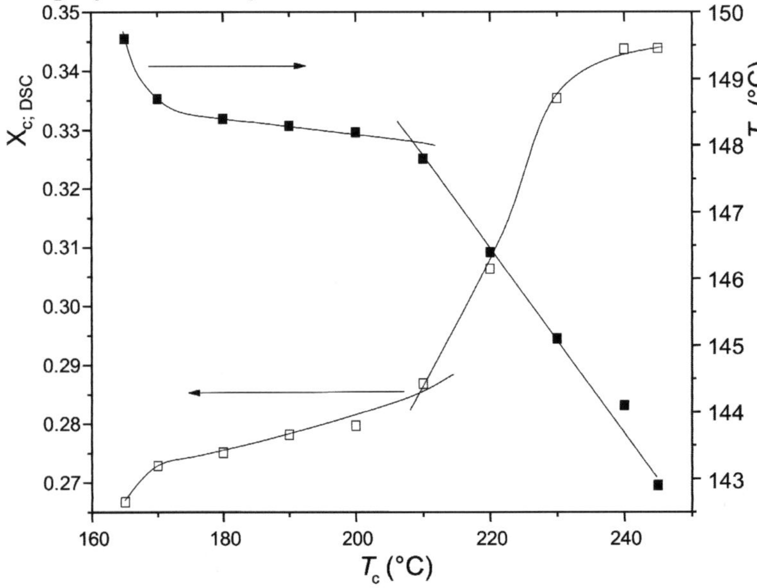

Figure 2: Mass crystallinity ratio values (□) as determined from DSC (X_{cDSC}) and α-relaxation temperatures T_α at 1950 Hz (■) *vs.* the crystallization temperature (T_c) for samples crystallized during t_c = 30 min

3.2. Relaxation Dynamics and nano-scaled morphology

The relaxation phenomena in amorphous and semi-crystalline PEN have been previously investigated by means of dynamic mechanical analysis (DMA) and dielectric relaxation spectroscopy (DRS) [4-7]. Three relaxation processes: β, β* and α have been observed in order of increasing temperatures:
- The β-relaxation can be observed at around -20°C for a frequency of 1 kHz. It is similar to the sub-glass transition of PET and is associated to the localized motions involving the ester groups[3,7,34,35].
- The β*-relaxation, appearing at around 100°C for a frequency of 1 kHz is another sub-glass transition with cooperative feature. This process was assigned to the out of plane motions of the naphthalene cycle by Bellomo *et al.* [36] and more precisely to the motions within nano-scale aggregates of naphthalene rings motions [4,6,7,37,38]. In the frequency/temperature region of the *β** process, the contribution of motions within liquid-crystalline nanodomains are also reported after crystallization from the molten state[42] which can complicate the exact attribution of the β* process.

- The α-relaxation which appears at around 140°C for a frequency of 1 kHz is associated to the glass transition. Thus, this process is due to cooperative segmental motions in the (mobile) amorphous regions within the lamellar stacks of the spherulites [2,10,12,39,40] or in the amorphous zones between spherulites.

The dielectric behaviors of amorphous and cold-crystallized PEN samples are displayed in Figure where $Tan(\delta)=\varepsilon''/\varepsilon'$ is plotted as a function of the temperature for the films crystallized during 24h at 165 and 245°C. The three relaxations described above can be observed. In the crystallization conditions chosen for this study, the temperature location of the β and β^* relaxation peaks are not significantly affected by crystallization, but the value of *Tan*(δ) in the entire relaxation pattern is decreased as the cristallinity ratio is increased, *i.e.* the number of mobile dipoles decreases. For the amorphous sample, a cold crystallization peak (CC) occurring at around 175°C can be observed. The occurrence of a $Tan(\delta)$ maximum in the crystallization region is not connected to a specific relaxation process (the temperature at maximum is not affected by frequency), but to a gradual change in the α-relaxation peak shape induced by cold crystallization and resulting confinement of the amorphous phase by the crystalline phase. A magnification in the glass transition region is represented as an onset in Figure The effect of crystallization classically shifts the glass transition relaxation to higher temperatures and also contributes to a broadening of the dielectric relaxation peaks.

A more systematic evolution of the α-relaxation temperature T_α at 1950 Hz is shown in Figure 3 (filled squares) displaying the decrease of T_α with crystallization temperature T_c. Again, a transition between two confinement regimes can be observed below and above the crystallization temperature 210°C. Theses relaxation results are consistent with the decrease of the calorimetric glass transition temperature with T_c as shown in Figure 1. As a conclusion, the apparent decrease of the glass transition relaxation and calorimetric glass transition temperatures with increasing crystallinity ratios and crystallization temperatures should be connected to the variation of the crystalline morphology and/or the structural state within the amorphous phase with T_c. Further analysis is needed to interpret the confinement effects of the crystalline onto the amorphous phase.

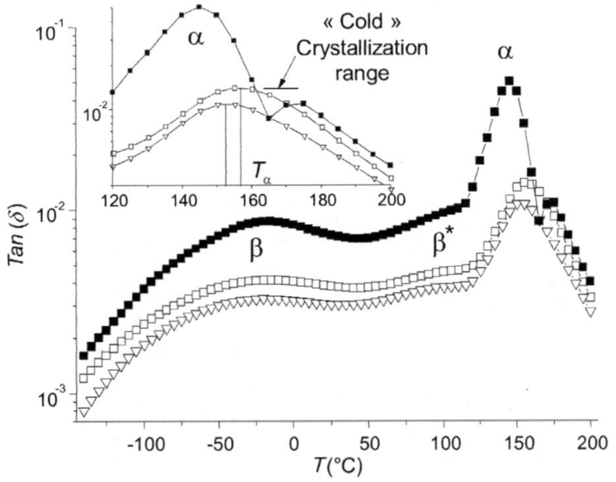

Figure 3: Tan(δ) as a function of temperature at 1950 Hz for amorphous PEN (■) and the samples crystallized during 24 hours at 165 (□) and 245°C (∇).

The fine structure of the lamellar stack morphology was studied by Small Angle X-ray Scattering. The values of the long period $L_p= \ell_a + \ell_c$ (deduced from the Bragg equation applied to the Lorentz-corrected SAXS data, see Figure 4), together with ℓ_a and ℓ_c thicknesses deduced from equation 5 and 6 are reported in Table 1 and Table 2

The long period L_p classically increases with crystallization temperature indicating that lamellar stacks with coarser morphology are developed for the high T_c values. In addition, L_p decreases while the annealing time increases. This could be related to the partial insertion of new lamellae (possibly originating from the slow secondary crystallization) between existing crystallites so that lamellar stacks with smaller average long period progressively develop[41]. Such morphology development significantly impacts the dynamics as a result of the additional confinement effects of this new crystalline population.

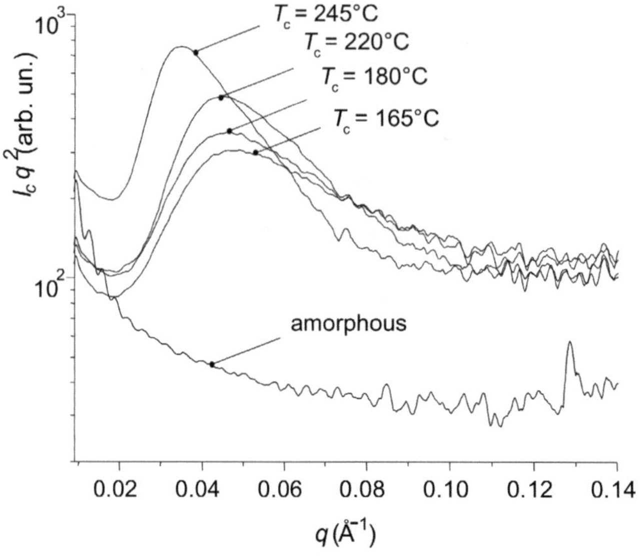

Figure 4: SAXS diagrams for the amorphous PEN and the samples crystallized during 2 hours at different temperatures.

T_c (°C)	t_c (h)	L_p (Å) (±3 Å)	ℓ_a (Å) (±3 Å)	ℓ_c (Å) (±2 Å)
	0.5	134	100	34
165	2	132	98	33
	24	125	93	32
	0.5	138	102	36
180	2	135	99	36
	24	124	88	36
	0.5	153	109	44
220	2	140	96	44
	24	135	91	44
	0.5	177	119	58
245	2	180	118	62
	24	174	108	66

Table 1: Lamellar stack morphology as determined by SAXS for cold-crystallized samples of PEN as a function of annealing time and temperature.

T_c (°C)	L_p(Å) (±3 Å)	ℓ_a (Å) (±3 Å)	ℓ_c (Å) (±2 Å)
170	135	100	35
180	138	102	36
190	142	104	37
200	145	106	39
210	147	107	39
220	153	108	44
230	153	109	49
240	169	113	56

Table 2: Lamellar stack morphology as determined by SAXS for the second set of the crystallized samples of PEN as a function of annealing temperature (t_c = 30 min)

4. Discussion

4.1 Geometrical analysis of confinement

The effect of crystallization on the "characteristic size" of the amorphous phase can first be discussed with the calculation of the amorphous lamellae thickness ℓ_a from SAXS diagrams (yielding L_p) and DSC (X_c) measurements from equation 6. Hence, assuming a localization of the rigid amorphous phase at the interface between crystal and amorphous lamellae, the central thickness $\ell_{a(\mathrm{MAF})}$ of the mobile amorphous fractions could be determined according to the following equation:

$$\ell_{a(\mathrm{MAF})} = (1\text{-}RAF_{DSC})^* \ \ell_a \qquad (5)$$

The corresponding calculations on the basis of Table 2 are ranging between 50 Å (T_c=180°C, t_c=24h) to 105Å (T_c =245°C, t_c=30min). Such characteristic widths of the MAF need to be compared to characteristic volume of cooperativity at the glass transition (ξ^3_{Tg}) calculated by using the Donth approach[41-43]:

$$\xi_{Tg}^3 = \frac{\Delta(1/C_v)}{\rho(\delta T)^2}.kT_g^2 \qquad (6)$$

where k is the Boltzmann constant, T_g is the glass transition temperature, $\Delta(1/C_v)$ is the step decrease in the reciprocal specific heat at T_g, ρ is the density of the amorphous material and $(\delta T)^2$ is the mean square temperature fluctuation related to the dynamic glass transition of a Cooperative Rearranging Region. $\Delta(1/C_v)$ can be evaluated by neglecting the difference between the heat capacity step at constant pressure and at constant volume:

$$\Delta(1/C_v) \approx \Delta(1/C_p) = (1/C_p)_{\mathrm{glass}} - (1/C_p)_{\mathrm{liquid}} \qquad (7)$$

Thus, for amorphous PEN, a characteristic volume of cooperativity of around 4.1 nm^3 and a characteristic size of around 16 Å (10 Å if a sphere volume is considered) were obtained. These values are significantly lower than the values obtained for the MAF thicknesses in semicrystalline systems exhibiting significant confinement effects. This difference is possibly related to the localization hypothesis: the rigid amorphous fraction may not be limited to the amorphous/crystal interface, but should be extended to any chain with conformation that is unfavourable for mobility, such as the

IOP Conf. Series: Materials Science and Engineering **48** (2013) 012005 doi:10.1088/1757-899X/48/1/012005

extended or stretched conformation between two adjacent crystallites. In this view, intercrystalline bridging chain segments or taut tie molecules could be considered to be part of the RAF even if they are not localized into the amorphous/crystal interface. Actually, such tie molecules should be considered as rigid entanglement obstacles for other amorphous chains, so that even if the tie-chain density is low, their effect on the RAF might be significant. The variation of the thermo-mechanical crystallization conditions should then be considered as a way to impact the semicrystalline morphology associated with a resulting tie-molecule density and their confinement effects.

In Figure 5(a) the α-process temperatures (at f = 1950 Hz) are plotted as a function of the respective values of ℓ_a. All the results are fairly rationalized in the $T_\alpha|_{f=1950\text{Hz}}$-$\ell_a$ plot, where the values of T_α appear largely impacted by the amorphous widths ℓ_a. It can be observed again that as the crystallization temperatures increase, a higher mobility is observed due to the increase of the width of the amorphous lamellae. Moreover, for the same crystallization temperature, as the annealing time increases, the molecular mobility decreases due to secondary crystallization inducing a lower value of ℓ_a. Yet, the morphology of the samples crystallized for a longer time (24 h) is related to the crystallization temperature in a complex way. The values of the final amorphous thickness appear to be minimal at a temperature close to 220°C (see Figure 5a, up triangle empty symbols), as a result of the combined effects of the initial lamellar stack morphology and the subsequent refinement of the microstructure induced by secondary crystallization as described by Strobl *et al*[44]. As a result, the values of ℓ_a do not seem to follow the same evolution for long crystallization times in comparison with the samples crystallized during 30 minutes where ℓ_a monotonously increases as the crystallization temperature increases (see Figure 5a, dark gray symbols and black symbols) but with a step-like evolution close to the crystallization temperature of 210°C.

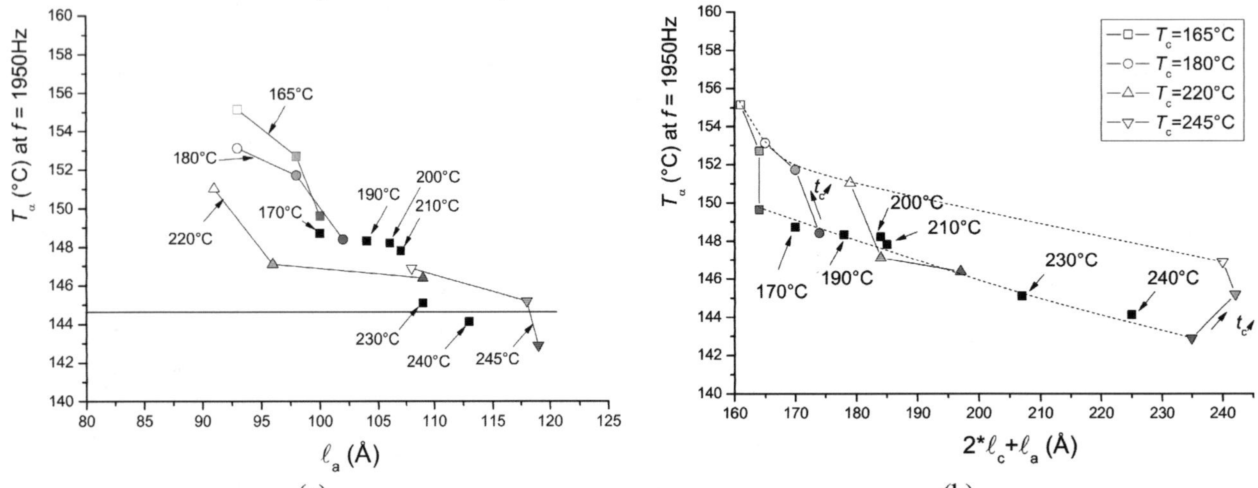

(a) (b)

Figure 5: (a) α-process temperatures (T_α) (± 0.5°C) as a function of the amorphous layer length (ℓ_a) for the samples crystallized at T_c = 165 (□), 180 (○), 220 (Δ) and 245°C (∇). The open symbols correspond to a crystallization time (t_c) of 24 h, the grey symbols to t_c = 2 h and the dark grey and black symbols (■) to t_c = 30 min. (b) α-process temperatures (T_α) (± 0.5°C) as a function of the characteristic length (2* ℓ_c +ℓ_a) (same symbols as in (a)).

4.2 Role of tie-molecules in the confinement effects.

Although a purely geometrical approach of confinement based on equations 5 and 6 does not yield alone a satisfactory frame of interpretation of the confinement effects in semicrystalline PEN, the evolution of the glass transition dynamics with semicrystalline morphology or crystallization temperature/time can be discussed invoking the role of tie molecules. In particular, the effect of the crystallization temperature can be first related to the tie molecule density through the degree of regular folding at the surface of the crystallites. Indeed, the works of Hoffman *et al.* [46-48] have shown that in the high crystallization temperature range, regular folding is facilitated by thermal activation, contrarily to lower crystallization temperatures. Thus, a higher number of tie molecules is expected to emerge from the crystals in order to connect the lamellar stacks in the samples crystallized at low temperatures. Another issue is the intrinsic effect of the crystal thickness and interlamellar spacing on the density of the tie molecules. Experimentally, the tie molecule density is known to decrease sharply with increasing long period L_p[49]. The relation between the tie molecule density, mean gyration radius of the chains and semicrystalline morphology parameters (L_p, ℓ_c, ℓ_a) was formalized in the theory of Huang and Brown[50-51] and illustrated in a variety of experimental investigations[52]. Accordingly, the increase of the characteristic length $2\ell_c+\ell_a$ decreases the probability to form a tie-chain molecule: a tie molecule is formed if the end-to-end distance of the random coil (before crystallization, in the melt) is larger than the critical length $2\ell_c+\ell_a$ (after the lamellar stacks are formed in the semicrystalline state). An illustration of the role of the characteristic length $2\ell_c+\ell_a$ is given in Figure 5(b), although longer crystallization/annealing times clearly yield to confinement laws slightly shifted to higher T_α values. This is indication that the semicrystalline morphology could be more complex than the description proposed in the frame of the lamellar stack model: the development of small non-lamellar crystals within the amorphous layers by secondary crystallization could have negligible impact on the mean long period as deduced by SAXS, but could significantly contribute to confinement effects by physical cross-linking. Such complex morphology could also explain the asymmetry towards the large q range tail of the SAXS patterns (see Figure 4). The effect of the crystallization temperature or annealing time was previously related to the RAF due to crystalline defects removal[53] in the case of PET: the decrease in the RAF is attributed to crystal perfection enhancement which in turn decreases the strain transmitted to the amorphous phase by means of the tie molecules[54]. In the present work, the increase of the crystallization temperature is indeed observed to induce a higher crystalline perfection, but it also leads to an increase of the crystallinity ratio (see Figure 2) that should increase the number of tie-chain molecule between the primary crystalline lamellae and the newly formed secondary crystalline entities. In the same way, increasing the annealing or crystallization time systematically induces an increase of the confinement effects as a result of crystallization and subsequent a decrease of L_p and ℓ_a, although no increase of the crystalline perfection could be evidenced from X-ray diffraction measurements. A rather unexpected result was observed for the samples crystallized during 30 minutes at 240 and 245°C. The α-process temperature (f = 1950 Hz) of these sample were comparable to the amorphous sample (see Figure 5b), in spite of the presence of the crystalline phase. This should be related to a lower entanglement density in the amorphous lamellae.

To go a step further, we have evaluated the optimal structural length ℓ^* connected to the glass transition dynamics, in the form of:

$$\ell^*=(\ell_a+k.\ \ell_c)/(1+k) \qquad (8)$$

where k was adjusted to find minimum dispersion around the general trend curve of the T_α-ℓ^* plot (for details of the numerical see supplementary material). The optimal result accounting for all the results is found for k=0.83 corresponding to an intermediate result between $\ell^*=\ell_a$ and $\ell^*=(\ell_a+2\ \ell_c)$ (see Figure 5a), but close to the value of $\ell^*=L_p$. Such critical length scaling with L_p was invoked in previous works[55,56] for the estimation of tie molecule density, which can be interpreted as an indication of the important role of tie molecule density in the confinement effects of the glass transition in semicrystalline polymers. The optimal plot is thus given in Figure 6. Here again, the high temperature crystallization data at T_c=245°C cannot be renormalized on the common pattern. In this temperature range, the evolution of T_α at 1950Hz should thus be governed by additional disentanglement processes that should be disconnected to semicrystalline morphology.

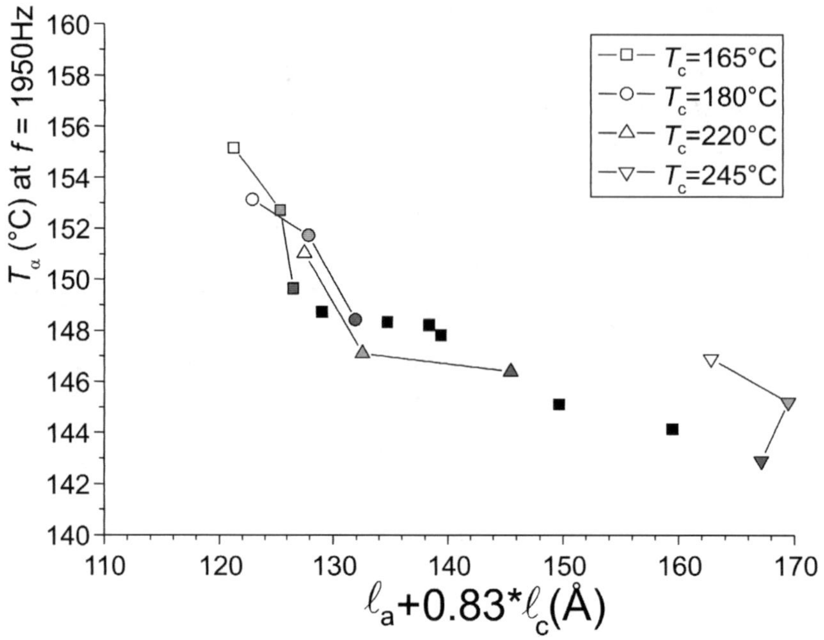

Figure 6: α-process temperatures (T_α) (± 0.5°C) as a function of the critical length $\ell^* = \ell_a + k\ \ell_c\ /(1+k)$ where k=0.83. The open symbols correspond to a crystallization time (t_c) of 24 h., the grey symbols to t_c=2 h and the dark grey and black symbols (■) to t_c = 30 min.

5. Conclusion

The effect of confinement of the amorphous phase by the crystalline nanophase in semi-crystalline PEN after crystallization in large temperature and crystallization time ranges has been studied. The evolution of the molecular mobility is shown to be largely impacted by the tie molecule density that is in turn related to the semicrystalline morphology. Such tie molecules should be considered as rigid entanglement obstacles for other amorphous chains, so that even if the tie-chain density is low, their effect on the glass transition dynamics and the RAF is significant. Accordingly, the characteristic length optimally describing the structural evolution of the dielectric glass transition relaxation was found to be $l^* = l_a + 0.83\ l_c \sim L_p$, where L_p is the long period, l_a is the inter-crystalline distance and l_c the crystalline thickness. At high crystallization temperatures (above 220°C) additional effects with the degradation of the polymer chains should be considered, leading to a decrease of the entanglement and tie molecule density and thus to lower confinement effects. Such analysis explains why the evolution of the frequency/temperature location of the dielectric α-relaxation peak should not be directly related to the crystallinity ratio but depends on the details of the crystalline lamellar stack morphology governing the tie-molecule density.

6. Acknowledgements

The authors thank Du Pont de Nemours (Lucien Schossler) for providing the PEN samples and Prof. Yves Grohens and Prof Jean-Pierre-Pascault for fruitful discussions. We are indebted to Cyrille Rochas and CRG group at ESRF for Synchrotron SAXS measurements. We are also grateful to Olivier Godard for DSC help in Institut Jean Lamour (Lorraine University).

7. References

[1] Huo P, Cebe P. *Macromolecules* 1992;**25**(2):902-909.
[2] Nogales A, Denchev Z, Ezquerra TA. *Macromolecules* 2000;**33**(25):9367-9375.
[3] Bravard SP, Boyd RH. *Macromolecules* 2003;**36**(3):741-748.
[4] Hardy L, Fritz A, Stevenson I, Boiteux G, Seytre G, Schönhals A. *Journal of Non-Crystalline Solids* 2002;**305**(1-3):174-182.
[5] Cañadas JC, Diego JA, Mudarra M., Belana J, Díaz-Calleja R, Sanchis MJ, Jaimés C. *Polymer* 1999;**40**(5):1181-1190.
[6] Hakme C, Stevenson I, David L, Boiteux G, Seytre G, Schönhals A. *Journal of Non-Crystalline Solids* 2005;**351**(33-36):2742-2752.
[7] Hakme C, Stevenson I, David L, Seytre G, Boiteux G. *Journal of Non-Crystalline Solids* 2006;**352**(42-49):4746-4752.
[8] Kalika DS, Krishnaswamy RK. *Macromolecules* 1993;**26**(16):4252-4261.
[9] Nogales A, Ezquerra TA, Denchev Z, Sics I., Calleja FJB, Hsiao BS. *The Journal of Chemical Physics* 2001;**115**(8):3804-3813.
[10] Ezquerra TA, Majszczyk J, Baltà-Calleja FJ, López-Cabarcos E, Gardner KH, Hsiao BS. *Physical Review B* 1994;**50**(9):6023.
[11] Wübenhorst M. *Polymeric Materials: Science & Engineering* 2007;**97**:829.
[12] Ezquerra TA, Sics I, Nogales A, Denchev Z, Balta-Calleja FJ. *Europhysics Letters* 2002; **59**(3):417-422.
[13] Pieruccini M, Ezquerra TA, Lanza M. *The Journal of Chemical Physics* 2007;**127** (10):104903-8.
[14] Kortaberria G, Marieta C, Jimeno A, Arruti P, Mondragon I. *Journal of Microscopy* 2006;**224**(3):277-289.
[15] Bras AR, Viciosa MT, Wang Y, Dionisio M, Mano JF. *Macromolecules* 2006;**39**(19): 6513-6520.
[16] Fitz BD, Andjelic S. *Polymer* 2003;**44**(10):3031-3036.
[17] Mijovic J, Sy JW. *Macromolecules* 2002;**35**(16):6370-6376.
[18] Fitz BD, Jamiolkowski DD, Andjelic S. *Macromolecules* 2002;**35**(15):5869-5872.
[19] Sics I., Ezquerra TA, Nogales A, Denchev Z, Alvarez C, Funari SS. *Polymer* 2003; **44**(4):1045-1049.
[20] Cheng SZD, Wunderlich B. *Macromolecules* 1988;**21**(3):789-797.
[21] Wunderlich B. *Progress in Polymer Science* 2003;**28**(3):383-450.
[22] Pierruccini M, Flores A, Nöchel U., Di Marco G, Stribeck N, Balta-Calleja FJ. *European Physical Journal E.* 2008;**27**:365-373.
[23] Puente-Orench I, Stribeck N, Ania F, Baer E, Hiltner A. Balta-Calleja FJ. *Polymer* 2009; **50**:2680-2687.
[24] Stewart ME, Cox AJ, Naylor DM. *Polymer* 1993; **34**(19):4060-4067.
[25] Nakamae K, Nishino T, Gotoh Y. *Polymer* 1995; **36**(7):1401-1405.
[26] Van den Heuvel CJM, KlopEA. *Polymer* 2000;**41**(11):4249-4266.
[27] Weick BL, Bhushan B. *Wear* 1995;**190**(1):28-43.
[28] Weick BL, Bhushan B. *IEEE Transactions on Magnetics* 1995; **31**(6):2937-2939.
[29] Serghei A, Kremer F, *Macromol. Chem. Phys.*2008;**209**:810.
[30] Ngai KL. *Comments on Solid State Physics* 1979;**9**:127-140.
[31] Ngai KL, White CT. *Physical Review B* 1979;**20**:2475-2486.
[32] Buchner S, Wiswe D, Zachmann HG. *Polymer* 1989;**30**(3):480-488.
[33] Patcheak T, Jabarin S *Polymer* 2001;**42**(21): 8975-8985.
[34] Buchner S, Wiswe D, Zachmann HG. *Polymer* 1989**;30**(3):480-488.
[35] Patcheak T, Jabarin S *Polymer* 2001;**42**(21): 8975-8985.
[36] Mencik Z. *Chemicky Prümysil* 1967**;** **17**(42):78.

[37] Maxwell AS, Monnerie L, Ward IM. *Polymer* 1998;**39**(26):6851-6859.
[38] Bellomo JP, Lebey T. *Journal of Physics D: Applied Physics* 1996;**29**(7):2052-2056.
[39] 40.Hardy L,Stevenson I., Boiteux G, Seytre G, Schönhals A. *Polymer* 2001;**42**(13):5679-5687.
[40] 41.Hardy L, Stevenson I., Fritz A, Boiteux G, Seytre G, Schönhals A. *Polymer* 2003;**44** (15):4311-4323.
[41] Hakme C , Stevenson I , David L, Voice A, Seytre G *Journal of Non-Crystalline Solids* 2007;**353**: 4252–4261.
[42] Coburn JC, Boyd RH. Macromolecules 1986;**19**(8):2238-2245.
[43] Roland CM, Ngai KL *Journal of non crystalline solids* 1997;**212**:74-76
[44] Strobl G; *in* The Physics of Polymers, Concepts understanding their structures and behavior, Springer, Berlin, 2nd Edition, 1997, ISNB 3-540-63203-4, p. 181
[45] Botev M, Neffati R, Rault J. *Polymer* 1999;**40**(18):5227.
[46] Menczel JD *Journal of Thermal Analysis and Calorimetry* 2011, pages1-18, Doi: 10.1007/s10973-011-1496-7
[47] Donth E. *Journal of Non-Crystalline Solids* 1982;**53**:325.
[48] Donth E. *Journal of Non-Crystalline Solids* 1996;**34**(17):2881-2892.
[49] Donth E. *Journal of Non-Crystalline Solids* 1999;**50**(7):240-251.
[50] Hoffman JD. *Polymer* 1982;**23**:656-670.
[51] Hoffman JD. *Polymer* 1983;**24**:3-26.
[52] Hoffman JD, Miller RL. *Polymer* 1997;**38**(13):3151-3212.
[53] Janimak JJ, Stevens GC *Journal of Materials Science* 2001;**36**(8):1879-1884
[54] Huang Y-L, Brown N *Journal of Materials Science* 1988;**23**(10):3648-3655
[55] Huang Y, Brown NJ *Polym. Sci., Part B: Phys* 1991;**29**:129
[56] Brown N, Zhou Z *Macromolecules* 1995;**28**:1807-1811

IOP Conf. Series: Materials Science and Engineering **48** (2013) 012006 doi:10.1088/1757-899X/48/1/012006

Influence of the mechanical stress and the filler content on the hydrostatic compression behaviour of natural rubber

Jan Zimmermann, Markus Stommel

Chair of Polymer Materials, Saarland University, Saarbrücken, Germany

j.zimmermann@mx.uni-saarland.de

Abstract. The behaviour of natural rubber (NR) compounds under mechanical stress is often reported in literature. An important and widely discussed effect that occurs is the Mullins effect. During the first loading cycles in a tensile test for example, a stress-softening effect is observed. This and other effects on the mechanical behaviour are investigated for different rubber materials with and without different types of fillers and filler contents. Besides, the hydrostatic compression behaviour is affected by the type and content of filler as well, which is shown for an NR with and without waxes and different contents of carbon black (CB) in this contribution. In contrast to the Mullins effect, there is no dependence of the number of loading cycles on the volumetric behaviour determined in hydrostatic compression tests. Furthermore, the influence of the previous stress-softening due to mechanical stress on the compression behaviour is elaborated. Cyclic uniaxial tensile tests are performed to realize the stress-softening in the rubber materials. The subsequent compression tests are compared to compression tests without any pre-stretching to determine the influence of previous mechanical loading on the compression behaviour of natural rubber with different filler contents.

1. Introduction

The dimensioning of rubber components by finite element (FE) simulations is generally applied. In common FE tools several material models are implemented to represent the hyperelastic material behaviour of rubber via a formulation of the strain energy function, for example Mooney [1], Ogden [2], Kilian [3] or Heinrich [4]. In general, the material behaviour and therewith the strain energy function is split into a volume preserving (deviatoric) and a volume changing (volumetric) part. Both parts can be expressed dependent on the invariants I_i of the left Cauchy-Green strain tensor, in which the first and second invariant describe the deviatoric, the third one the volumetric part of the respective function. The derivation of these functions results in an expression for the stresses. The approach according to Kilian [3] is stated in the equation (1) as well as the respective Cauchy stress:

$$W = \underbrace{\mu\left\{-\left(\lambda_m^{\;2}-3\right)\left[\ln(1-\eta)+\eta\right]-2/3\alpha\left(\frac{\tilde{I}-3}{2}\right)^{\frac{3}{2}}\right\}}_{W_{dev}(I_1,I_2)} + \underbrace{\frac{K}{2}\left(\frac{J_{el}^{\;2}-1}{2}-\ln J_{el}\right)}_{W_{vol}(I_3=J_{el})} \tag{1}$$

with

$$\eta = \sqrt{\frac{\tilde{I} - 3}{\lambda_m^{\ 2} - 3}} \tag{2}$$

$$\tilde{I} = (1-\beta)\bar{I}_1 + \beta \bar{I}_2 \tag{3}$$

and

$$\bar{I}_1 = \bar{\lambda}_1^{\ 2} + \bar{\lambda}_2^{\ 2} + \bar{\lambda}_3^{\ 2} \tag{4}$$

$$\bar{I}_2 = \bar{\lambda}_1^{\ 2}\bar{\lambda}_2^{\ 2} + \bar{\lambda}_1^{\ 2}\bar{\lambda}_3^{\ 2} + \bar{\lambda}_2^{\ 2}\bar{\lambda}_3^{\ 2} \tag{5}$$

$$\bar{\lambda}_i = J^{-1/3}\lambda_i \tag{6}$$

$$J_{el} = \det \mathbf{F} = \lambda_1^{\ 2}\lambda_2^{\ 2}\lambda_3^{\ 2} \tag{7}$$

$$\boldsymbol{\sigma} = \frac{1}{J}\mathbf{B} \cdot \frac{\partial W}{\partial \mathbf{B}} = \mathbf{S} - p\mathbf{I} \tag{8}$$

μ = initial shear modulus; λ_m = locking stretch; α = global interaction parameter; β linear mixture parameter; $\bar{I}_i$ = invariants of the left Cauchy-Green strain tensor B; J_{el} = elastic volume strain; K = bulk modulus; λ_i = (deviatoric) stretch; F = deformation gradient; σ = Cauchy stress tensor

The required material parameters for the deviatoric part are determined by one or a combination of different mechanical tests like uniaxial tensile or simple shear testing. The assumption of nearly incompressibility and therefore the precondition of the bulk modulus K exceeding the material's shear modulus μ by several orders of magnitude is commonly accepted and applied [5]. Concerning simple loading conditions of a rubber part, a material model calibration based on these tests and the precondition K/μ >> 1 provides a sufficient result quality of FE simulations. Hereby, K is set based on experience in common practice, since its impact on the FE result is nearly negligible [6]. The influence of the volumetric part and therefore its material parameter, the bulk modulus, increases significantly when rubber components are (partially) exposed to hydrostatic pressure, like for highly confined rubber components where a change in shape of the rubber is constrained [6]. Several approaches were published with the aim to further improve the simulation quality with respect to the volumetric part of the strain energy function of which some are summarized by Hartmann and Neff [7]. The material parameter related to the volumetric part, the bulk modulus K is defined in equation (9):

$$K = \frac{\partial p}{\partial(\Delta V / V_0)} \tag{9}$$

p = pressure; ΔV = volume reduction; V_0 = initial volume

The mechanical behaviour and therewith the material parameters like shear modulus is highly affected by the rubber's composition or more precisely by the type and content of its fillers. Carbon black (CB) as active filler increases the material's stiffness as well as its wear hardness for example. Ageing inhibition and plasticizer are added to improve the component's performance. The material modelling and simulation are based on versatile mechanical tests of rubber compounds to determine their behaviour under different loading conditions. During the first loading cycles a stress-softening occurs which is known as Mullins effect [8]. Due to partially irreversible processes like breakage of filler agglomerates the stress softening as well as a permanent set within the first four to ten repetitions of a quasi-static loading and unloading is observed. Since the first description of this effect by Mullins its cause or the influence of the type of rubber and filler are investigated in many publications and are still widely discussed. [8-14]

In contrast to effects related to the mechanical loading condition and its history the hydrostatic compression behaviour in general and the influence of fillers on this in particular are rarely or not at all reported in literature to the author's knowledge. In [5] and [15] the dependence of relative volume reduction on hydrostatic pressure is described for a filled, technical natural rubber (NR) compound which shows the same characteristics as depicted in Figure 1. The bulk modulus corresponds to the ratio of Δp and $\Delta V/V_0$ and reaches a nearly constant value when the volume strain exceeds a certain level.

In this contribution the volumetric behaviour for a technical NR compound is determined. Further hydrostatic compression tests are conducted to evaluate the influence of decreasing carbon black content as well as the impact of omitted additives like plasticisers or antioxidants. The compounds are exposed to four loading unloading cycles to determine a possible effect on the volumetric behaviour. Based on these test results the evolution of the ratio K/μ in dependence of the CB content is analysed, which represents a crucial parameter for achieving an adequate simulation result quality. Due to the stress softening effect occurring during the first loading cycles, the influence of uniaxial pre-stretching on the hydrostatic compression behaviour is investigated.

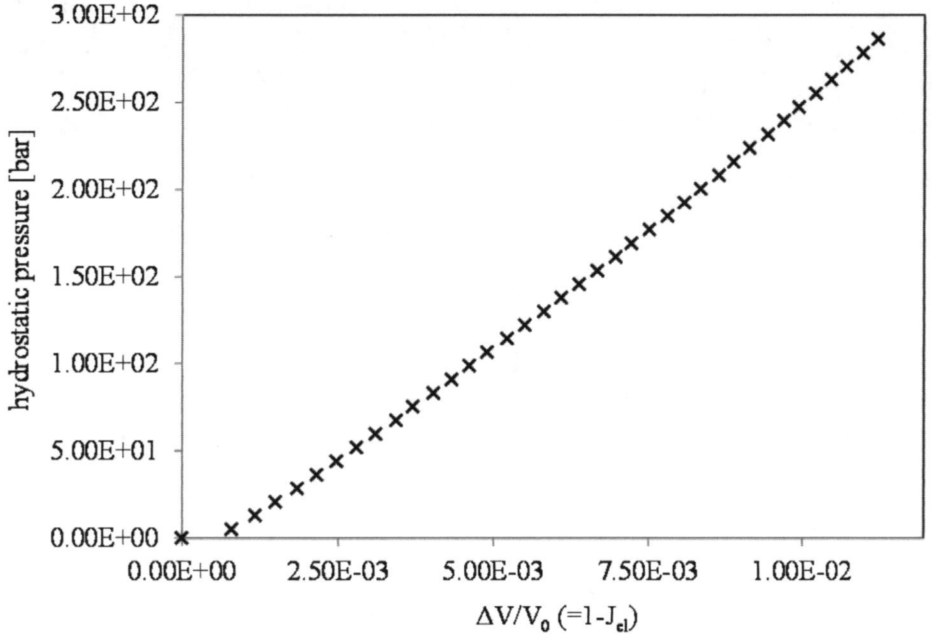

Figure 1. Relation of hydrostatic pressure on relative volume reduction [6].

2. Experiments

Cylindrical samples were prepared with a technical NR compound used for tread surfaces in automotive industry (I) which is shown in Table 1. Based on this compound the CB content in II to IV is reduced by one third each, whereas in V to VIII plasticiser and ageing inhibition, columns 4 to 7 in Table 1, is left out though they show the same CB content as I to IV, respectively. Compound VIII represents an unfilled NR. I-IV are named "with waxes", V-VIII "without waxes". For each compound three hydrostatic compression tests are conducted with transfer moulded, cylindrical specimens (diameter: 18 mm, length: 100 mm). To extract and eliminate the system's response a reference measure is conducted using a steel specimen instead of rubber with the same volume as the samples. A more detailed description of the testing procedure can be found in [6].

The mean value and standard deviation of three compression tests results for each compound are evaluated. The first of the three specimens is exposed to four loading cycles to detect a possible influence of previous loading steps during compression tests. Another set of hydrostatic compression tests is conducted to investigate the influence of stress-softening due to uniaxial tension on the resulting volumetric behaviour under hydrostatic pressure. Therefore rubber plates of each compound were prepared and subjected to four cyclic uniaxial loadings. In a second step these plates were cut into stripes of the same volume as the cylindrical samples to carry out hydrostatic compression tests. Furthermore the stress strain behaviour during uniaxial loading is recorded and analysed to obtain a value for the initial shear modulus for each compound. Therewith the evolution of K/μ depending on the filler content can be gained.

Table 1. Composition of investigated rubber compounds.

Ingredient	I	II	III	IV	V	VI	VII	VIII
					[phr]			
NR					100			
carbon black N115	50	33	17	-	50	33	17	-
zinc oxide					4			
Vulkanox 4020 (6PPD)		2					-	
Vulkanox HS (TMQ)		2					-	
Antilux 654		1,5					-	
Vivatec 500		4					-	
stearic acid					2			
sulphur					1,75			
Vulkacit CZ (CBS)					1,75			

3. Results and discussion

In this chapter the results of the conducted experiments are presented and interpreted. The effects occurring are compared to findings published so far. First, results of simple hydrostatic compression test are presented, followed by those of pre-stretched compression tests. Afterwards, the impact on FE simulations is discussed briefly. Furthermore, the dependence of the number of loading cycles on the volumetric behaviour is determined.

3.1. Hydrostatic compression tests

The influence of CB content as the main reinforcing additive in NR compounds on the volumetric behaviour is determined by hydrostatic compression tests. In Figure 2 the results for compound I to IV (s. Table 1) are presented, the standard deviation recorded for the different tests is less than 3 %. For lower volume strain values the pressure response shows a slight curvature where the gradient of the curve and therefore the bulk modulus increases about one decade. The response obtained after exceeding a pressure level of about 10 bar is nearly linear and represented by a constant bulk modulus. Since there is some gas in the system, the very first volume reduction compresses this and does not affect a rise in pressure, which is considered in the evaluation and the resulting curves are shifted horizontally towards the origin. This characteristic curvature may be caused by atomic-scale free volume holes, which are for example studied in (16, 17). When volume inside the chamber is reduced, at the beginning, the free volume impedes a soaring of pressure until it is occupied by water or the holes are nearly closed. The lower resistance against compression leads to the curvature at small volume strain.

An increasing CB content effects a decrease in the number of free volume holes [16] and therefore reduces the range of volume strain with a non-linear pressure response and lower bulk modulus (s. Table 2). Furthermore, by increasing the CB content, the gradient, and therefore the bulk modulus increase as well. The pressure response of compound IV deviates from the other curves, a significant increase in pressure is first detected for volume strain exceeding 0.01. This may be due to trapped air inside the specimen during the curing process as well as a larger number and size of free volume holes, since sulphur leads to these both effects and this compound does not possess any carbon black which reduces this number. A similar result is obtained for compounds without plasticiser and ageing inhibition, V to VIII, which show a curvature for lower pressure values followed by a linear progression for higher values (s. Figure 3).

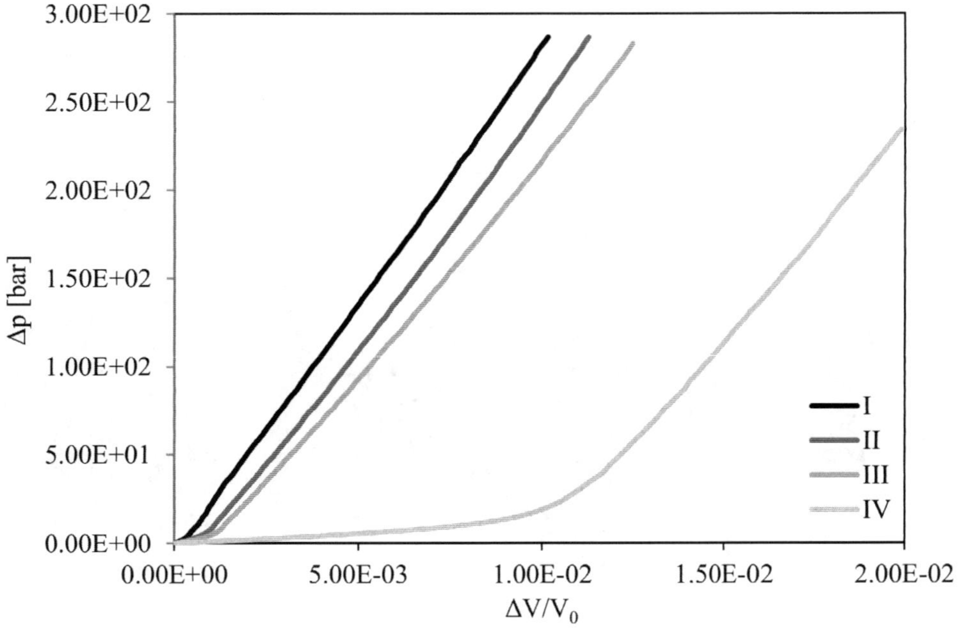

Figure 2. Influence of CB content on hydrostatic compression behaviour (I-IV).

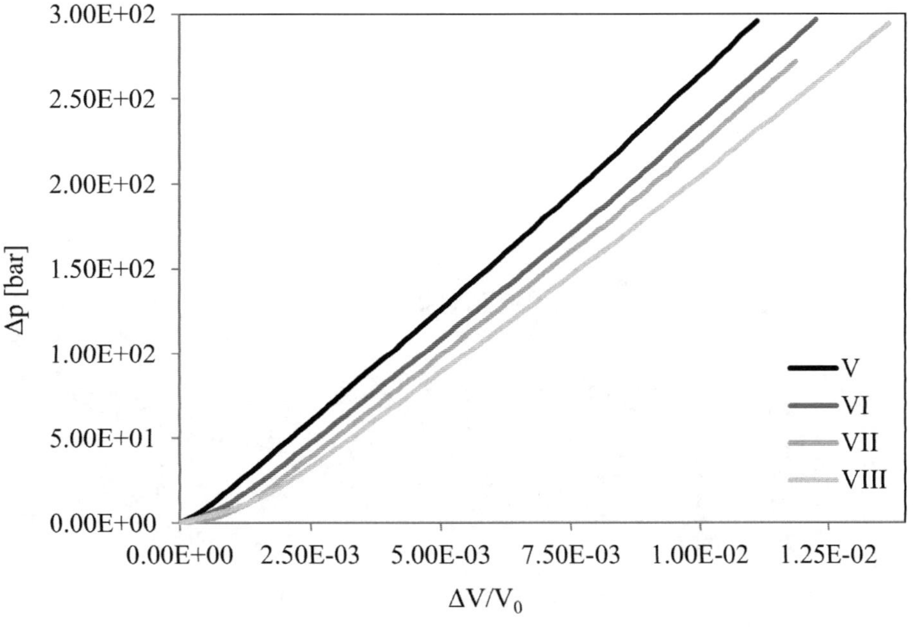

Figure 3. Influence of CB content on hydrostatic compression behaviour (V-VIII).

The resulting bulk moduli for the linear range of the pressure response are summarized in Table 2. A reduction of the CB content leads to a decrease in bulk modulus for both NR with and without waxes of approximately 20 %). Furthermore, compounds without waxes show about 4 % higher bulk moduli compared to those with waxes for the same CB content, respectively, which leads to the conclusion that the bulk stiffness is increased for those rubbers. This effect is due to the higher volume fraction occupied by CB compared to technical NR, since the bulk modulus of carbon black exceeds those of rubber and waxes by an order of magnitude ($K_{CB} \approx 30$ GPa). On the other hand waxes reduce the bulk modulus of NR which can be seen by comparing both compounds without any CB. This effect may be reasoned in the higher compressibility of waxes compared to NR as well as in the filler-rubber network which is built during in the curing process. A more detailed determination of the observed effects and its cause is topic of current research.

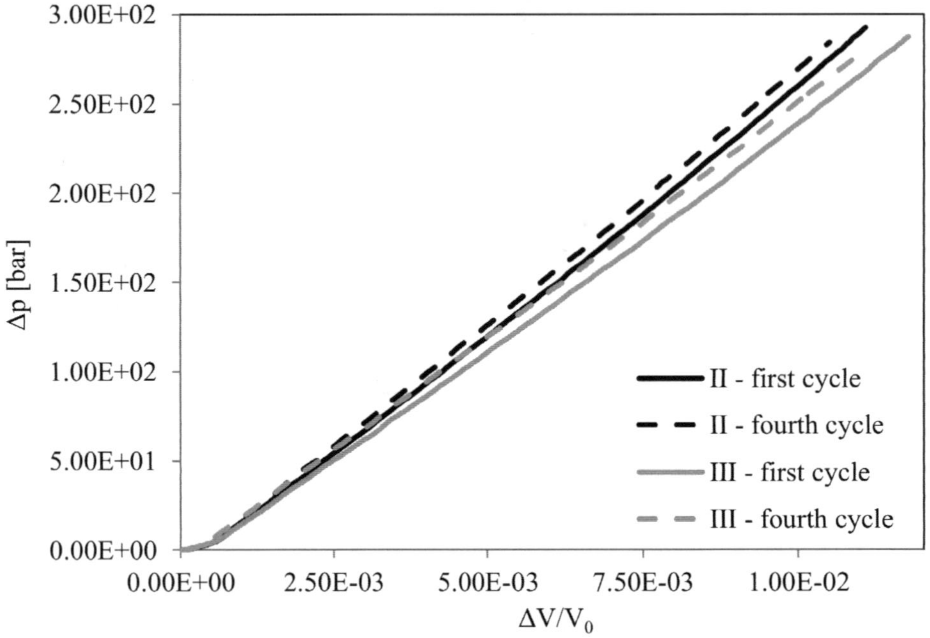

Figure 4. Influence of loading cycles (II and III).

3.2. Hydrostatic compression tests with previous uniaxial tensile loading

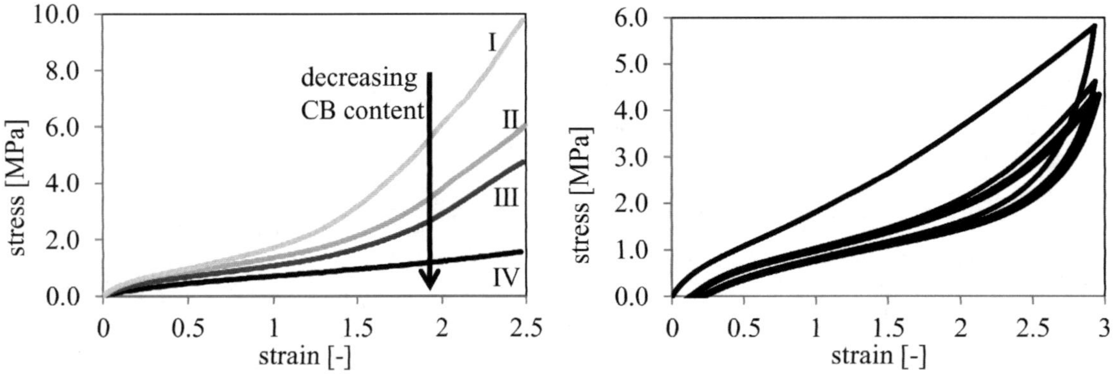

Figure 5. Uniaxial tensile tests - influence of CB content (left), stress-softening effect (right).

Besides the initial hydrostatic compression behaviour as shown in Figure 2 and Figure 3 the effect of further loading cycles is determined. In Figure 4 the resulting pressure over volume reduction of the first and fourth cycle is depicted for compounds II and III representing a technical rubber compound with 33 phr CB or 17 phr CB, respectively. The corresponding curves still possess a similar progression whereas the hydrostatic pressure with respect to volume strain, and therefore the bulk modulus, shows an increase from the first to the last loading cycle of about 10 % for technical NR compounds with 0 phr or 17 phr CB, respectively. This phenomenon is caused by free volume holes as well. During each compression phase, free volume is compressed and partially occupied by penetrating water. This will not be removed completely before the following loading step, so that the number and size of free volume holes decreases which reflects in the slightly higher increase in curvature at low volume strain with on-going loading. Furthermore, an offset in volume, that is necessary to arouse an increase in pressure, is detected, which decreases with the number of loading cycles. A possible breakage of filler cluster may also contribute to the observed behaviour. Besides, due to the occupied volume the increase in bulk modulus can be detected. The compounds with higher CB content show a slightly lower increase in K of about 6 %, since the number of free volume holes reduces with higher CB content [16]. A continuing testing after the first four loading cycles leads to a steady state in the dependence of volume strain and hydrostatic pressure.

The compound without waxes show the same behaviour, there is an increase in bulk stiffness detected during the very first loading cycles. In future research swelling experiments are to be conducted to determine the absorption of each specimen and more precisely each compound. A comparison of these results to the volume reduction found during the cyclic compression tests is to be drawn to confirm the assumption of free volume holes influencing the hydrostatic behaviour.

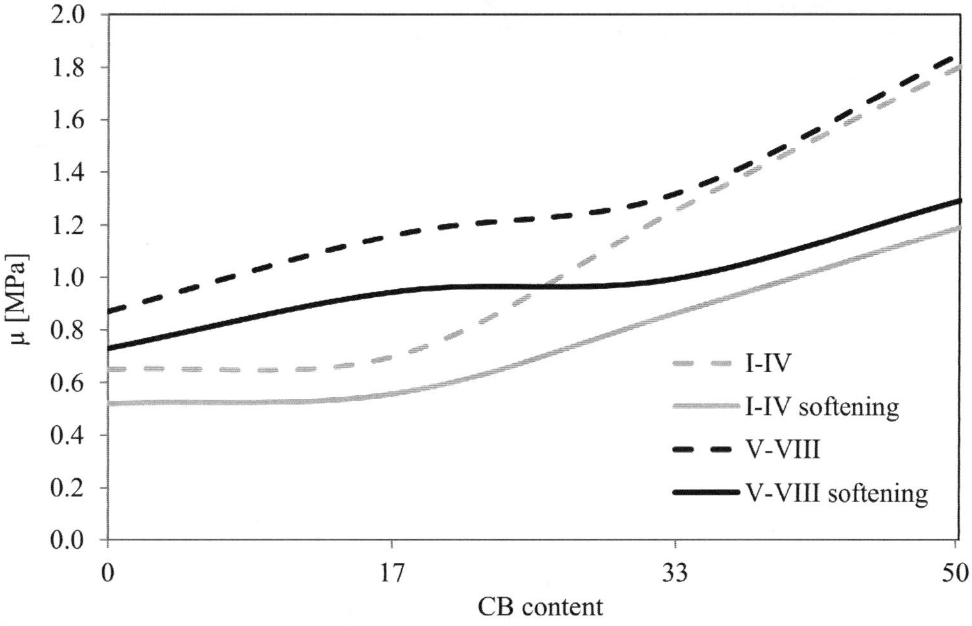

Figure 6. Influence of filler content and stress-softening on shear modulus.

Since the stress-softening effect significantly influences the mechanical behaviour and directly the shear modulus, a possible effect of uniaxial pre-stretching on hydrostatic compression is investigated. Besides, for modelling the material behaviour in FE analysis the shear modulus as a material parameter is required. Therefore uniaxial tensile tests are conducted and the stress strain response is extracted. Figure 5 shows exemplary decreasing stress responses due to a decreasing CB content (I-IV) as well as the stress-softening effect for compound III. The evolution of μ with respect to filler content as well as a comparison of the resulting shear modulus with and without stress-softening is depicted in Figure 5. The stress-softening effect occurs even for compounds without any CB, however the gap between μ with and without softening increases with CB content. Moreover, an increasing CB content causes an increasing shear modulus for both, compounds with and without waxes, respectively. Waxes soften the material and therefore lead to a lower μ (I-IV) compared to compounds without waxes. This effect becomes less significant for high CB content (greater than 33 phr), which can be seen by the developing of the respective curves in Figure 6. Therein, the volume fraction of CB is high compared to those of waxes so that the reinforcement clearly dominates the material's behaviour.

The influence of pre-stretching on the volumetric behaviour is determined by exposing the specimen used in uniaxial tensile testing to hydrostatic pressure. The testing procedure is the same as in 3.1., whereas the specimens' shape differs since they are cut out of plates which were exposed to uniaxial loading conditions to introduce the stress-softening effect. The results of this test series is shown in Table 2, the compounds with 33 phr CB are left out here. The influence of pre-stretching on the compression behaviour of a technical rubber compound (I to IV) is negligible, since the results lie within the standard deviation. On the other hand, the pre-stretching influences the compression behaviour of compounds without waxes which results in a decrease in bulk modulus of about 5 %. This may be due to the breakage of filler network and therefore a lower resistance against external loading.

Table 2. Resulting bulk moduli with and without pre-stretching.

compound	bulk modulus [GPa]	
	without pre-stretching	with pre-stretching
I	2.68	2.65
II	2.50	-
III	2.30	2.28
IV	2.17	2.17
V	2.79	2.69
VI	2.58	-
VII	2.42	2.23
VIII	2.26	2.13

3.3. Comparison of shear modulus and bulk modulus

The results obtained for hydrostatic compression tests as well as shear tests enable a determination of the ratio of the respective material parameters K and μ, which is crucial for the result quality of FE simulations (6). Since the bulk modulus affects these results significantly for pressure values greater than 10 bar, the linear approximation of K is evaluated and compared to μ. The initial shear modulus as well as the bulk modulus is obtained for each of the eight NR compounds. In Figure 7 the resulting K/μ is depicted for technical NR compounds (I to IV) with decreasing content of carbon black.

The commonly made assumption of a bulk modulus being orders of magnitude higher than the shear respective modulus is confirmed, even though the ratio decreases with increasing content of CB. The higher the content of CB and therefore the reinforcement the more important becomes K in relation to μ, the ratio reduces to approximately two thirds compared to the NR without any CB. The shear stiffness increase under mechanical loading due to higher reinforcing filler content exceeds the increase in bulk stiffness. The assumption of a constant ratio of K and μ independent of the filler content may lead to a significant deviation in FE results and therefore influences the accordance of simulation and test results [6]. This instance and therefore the not constant ratio emphasize the importance of a determination of all material parameters including the bulk modulus to achieve a satisfactory FE result quality, especially for confined rubber components.

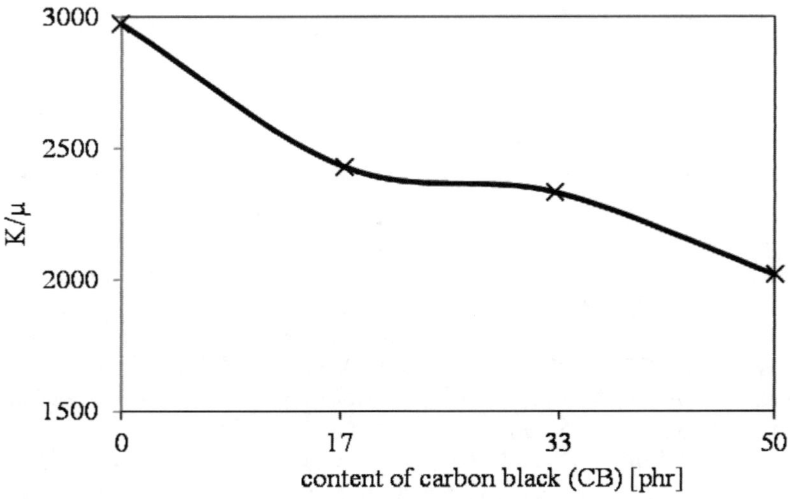

Figure 7. Influence of CB content on ratio K/μ (I-IV).

4. Conclusion

The results presented show the influence of fillers in NR, carbon black and waxes like ageing inhibition or plasticisers, on the hydrostatic compression behaviour. Therefore, NR compounds with four different carbon black contents each with and without waxes are tested. All of the tested compounds possess the same characteristic dependence of hydrostatic pressure on volume reduction. For low volume strain a curvature results from free volume holes, which show lower resistance against compression compared to the bulk material and are therefore compressed primarily, ending in a linear curve progression at higher volume strain or pressure values greater than 10 bar, respectively. A reinforcement as known from mechanical testing is found for hydrostatic compression as well. Due to a higher bulk modulus of CB compared to NR and a decreasing number of free volume holes in compounds with higher CB content the bulk stiffness and therewith its modulus increases. The effects on shear stiffness proved more significantly compared to bulk stiffness. The stress-softening for NR compounds under uniaxial loading as well as its influence on the hydrostatic compression behaviour is determined. It is pointed out, that there is no relation between pre-stretching and subsequent compression detected for technical NR compounds. Furthermore, a comparison of the development of shear and bulk modulus related to the content of CB is drawn. The influence of the volumetric part in material modelling becomes more important with an increasing CB content. Carbon black enhances the shear stiffness significantly compared to bulk stiffness.

To confirm that free volume holes cause the obtained behaviour under hydrostatic pressure, swelling experiments are to be carried out. Therewith, the free volume of a specimen and the influence of fillers on it can be determined. Furthermore, positron annihilation lifetime spectroscopy (PALS) proved to be an adequate tool for a more precise identification of the number and size of free volume holes and is to be conducted. A comparison of these tests and the here presented results will enable a more precise understanding and determination of the occurring effects in NR.

References

[1] Mooney M 1940 *A Theory of Large Elastic Deformation* Journal of Applied Physics 11 p 11
[2] Ogden R W 1972 *Large Deformation Isotropic Elasticity - on the Correlation of Theory and Experiment for Incompressible Rubberlike Solids.* Royal Society of Applied Mathematics and Physics vol 326
[3] Kilian H G 1981 *Energy-Balance in Networks Simply Elongated at Constant Temperature* Colloid and Polymer Science 259 pp 1084-1091

[4] Heinrich G, and Kaliske M 1997 *Theoretical and Numerical Formulation of a Molecular Based Constitutive Tube-Model of Rubber Elasticity* Computational and Theoretical Polymer Science 7 pp 227-241
[5] Ogden R W 1984, *Non-Linear Elastic Deformations* (Chichester, Ellis Harwood Ltd.)
[6] Zimmermann J, and Stommel M 2012 *The Mechanical Behaviour of Rubber under Hydrostatic Compression and the Effect on the Results of Finite Element Analyses* Archive of Applied Mechanics 83 pp 293-302
[7] Hartmann S, and Neff P 2003 *Polyconvexity of Generalized Polynomial-Type Hyperelastic Strain Energy Functions for near-Incompressibility* International Journal of Solids and Structures 40 pp 2767-2791
[8] Mullins L 1947 *Effect of Stretching on the Properties of Rubber* Journal of Rubber Research 16
[9] Mullins L, and Tobin N R 1965 *Stress Softening in Rubber Vulcanizates. Part I. Use of a Strain Amplification Factor to Describe the Elastic Behavior of Filler-Reinforced Vulcanized Rubber* Journal of Applied Polymer Science 9 pp 2993-3009
[10] Harwood J A C, Mullins L, and Payne A R 1965 *Stress Softening in Natural Rubber Vulcanizates. Part Ii. Stress Softening Effects in Pure Gum and Filler Loaded Rubbers* Journal of Applied Polymer Science 9 pp 3011-3021
[11] Harwood J A C, and Payne A R 1966 *Stress Softening in Natural Rubber Vulcanizates. Part Iii. Carbon Black-Filled Vulcanizates* Journal of Applied Polymer Science 10 pp 315-324
[12] Harwood J A C, and Payne A R 1966 *Stress Softening in Natural Rubber Vulcanizates. Part Iv. Unfilled Vulcanizates* Journal of Applied Polymer Science 10 pp 1203-1211
[13] Harwood J A C, and Payne A R 1967 *Stress Softening in Natural Rubber Vulcanizates. Part V. The Anomalous Tensile Behavior of Natural Rubber* Journal of Applied Polymer Science 11 pp 1825-1834
[14] Kahraman H, Weinhold G W, Haberstroh E, and Itskov M 2010 *Anisotroper Mullins-Effekt bei rußgefüllten Elastomeren* KGK 3 pp 64-69
[15] Wood L A, and Martin G M 1964 *Compressibility of Natural Rubber at Pressures Below 500kg/Cm2* Journal of Research of the National Bureau of Standards Section a-Physics and Chemistry A 68 p 259
[16] Jobando V O, and Quarles C A 2007 *Effect of Cross-Linking on the Free Volume Properties of Natural Rubber* Physica Status Solidi C - Current Topics in Solid State Physics vol 4 Physica Status Solidi C-Current Topics in Solid State Physics 10 (A. P. Knights, P. Mascher and P. J. Simpson) pp 3759-3762

IOP Conf. Series: Materials Science and Engineering **48** (2013) 012007 doi:10.1088/1757-899X/48/1/012007

Development of constitutive model for composites exhibiting time dependent properties

L Pupure[1], R Joffe[1,2,3], J Varna[1] and B Nyström[2]
[1]Material Science, Luleå University of Technology, S-97187 Luleå, SWEDEN
[2]Materials and Manufacturing, Swerea SICOMP, S-94126 Piteå, SWEDEN

E-mail: liva.rozite@ltu.se, roberts.joffe@ltu.se, janis.varna@ltu.se, birgitha.nystrom@swerea.se

Abstract. Regenerated cellulose fibres and their composites exhibit highly nonlinear behaviour. The mechanical response of these materials can be successfully described by the model developed by Schapery for time-dependent materials. However, this model requires input parameters that are experimentally determined via large number of time-consuming tests on the studied composite material. If, for example, the volume fraction of fibres is changed we have a different material and new series of experiments on this new material are required. Therefore the ultimate objective of our studies is to develop model which determines the composite behaviour based on behaviour of constituents of the composite. This paper gives an overview of problems and difficulties, associated with development, implementation and verification of such model.

1. Introduction

Natural fibre composites exhibit good mechanical properties [1-3] but until now they have been used mainly for non-structural applications [4-6]. Only during the few past years the development of natural fibre composites for structural applications has gained the momentum. Most of the work in development of these materials was focused on the composites with synthetic thermosetting matrices [7-8]. Very recently several commercial bio-based thermosetting resins (Tribest, EpoBioX etc.) have been introduced, allowing production of whole bio-based composites. Properties of polymers derived from soybean oil and protein fillers have been reported in [9], a critical review of bio-based thermosets is presented in [10]. The latest results [11-12] demonstrated that these composites are comparable with glass fibre reinforced polymers in terms of stiffness. However due to variability of fibre properties [13-15] and limited filament length it is complicated to arrange and control fibre alignment in composites as well as ensure stable, predictable composite properties. Therefore, another type of reinforcement with natural origin has caught attention of researchers – Regenerated Cellulose Fibres (RCF). These fibres are continuous with constant, reproducible cross-section and properties (see Figure 1.) however they exhibit highly non-linear time-dependent behaviour. Thus, in order to accurately describe these materials and predict their properties RCF have to be treated as material with time-dependent properties. In the same way as synthetic fibres, RCF can be aligned and used in composite manufacturing. Also the transverse failure mechanisms are similar (Figure 1a) – crack propagation along the interface of the fibre and in the matrix is similar to glass fibre composites.

[3] To whom any correspondence should be addressed.

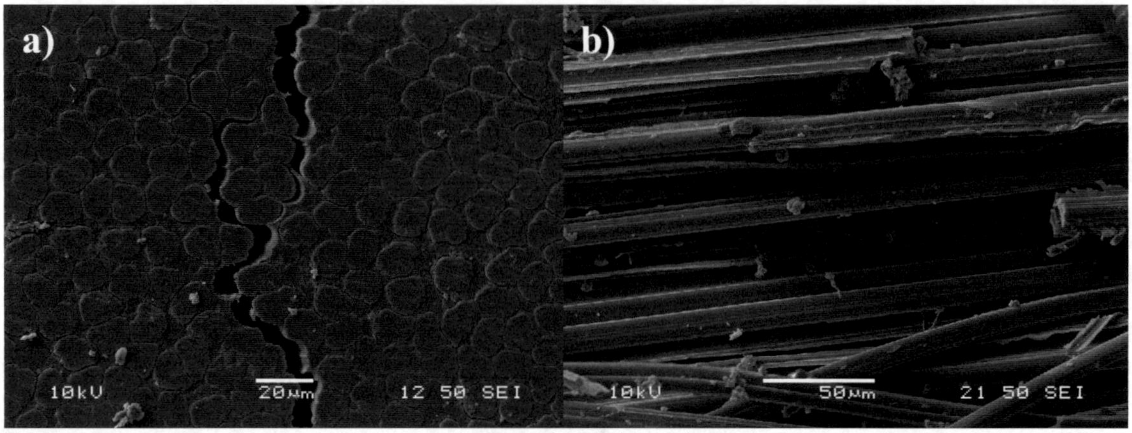

Figure 1. Scanning electron microscopy image of RCF/EpoBioX composite a) cross-section of fibers seen from the specimen edge and b) side view of fibers seen from the fracture surface

The nonlinear visco-elastic material model developed by Schapery [16-17] has been successfully applied to randomly oriented short bio-based composites [18-19] and synthetic long fibre composites [20]. However in order to apply this model, large number of time-consuming experiments must be carried out. As soon as the fibre orientation distribution or the fibre content changes new series of tests are required. Therefore it would be more convenient to use multiscale modelling, obtaining the higher scale material model on the basis of material models of constituents and their geometrical parameters in the composite. There are number of problems associated with development, implementation and verification of such model. These issues are discussed in this paper.

The used bio-based materials (resins and fibres) are relatively new and limited amount of information is available, especially when it comes to the long term performance. Most of the thermo-mechanical characteristics of these materials have to be obtained experimentally.

2. Model: Theory and experiments

2.1. Theoretical background

The theory of nonlinear viscoelastic materials used in this study was developed by Schapery [16-17], in this theory are obtained using expansion of the Gibb's free energy in viscoelasticity related internal state variables and using for them linear evolution laws with respect to thermodynamic forces. For anisotropic materials subjected to general loading case the resulting constitutive equations include many stress invariant dependent material parameters/functions to be experimentally identified. All these stress dependent functions are also affected by temperature and humidity, but in fixed environmental conditions they can be considered as stress dependent only.

In uniaxial tensile loading case, one-dimensional model [16-17] may be used, that contains only three stress invariant dependent functions, which characterize the nonlinearity. The model has been modified to account for microdamage $d(\sigma_{\max})$ [21]. The final form of the material model for one-dimensional case is as follows:

$$\varepsilon = d(\sigma_{\max})\left[\varepsilon_0 + g_1\int_0^t \Delta S(\psi - \psi')\frac{d(g_2\sigma)}{d\tau} + \varepsilon_{pl}(t,\sigma)\right] \quad (1)$$

The "reduced time" in equation (1) is given as:

$$\psi = \int_0^t \frac{dt'}{a_\sigma} \text{ and } \psi' = \int_0^\tau \frac{dt'}{a_\sigma} \quad (2)$$

In equation (1) ε_0 represents elastic strain in undamaged composite which, generally speaking, may be nonlinear function of stress. $\Delta S(\psi)$ is the transient component of the viscoelastic creep compliance at low stress, which characterize time dependent part of the Viscoelastic (VE) response. According to the theory it does not depend on stress level. The g_1 and g_2 are stress dependent material properties and a_σ is the shift factor. These three functions in fixed conditions depend on stress only.

It was shown by Schapery [17] that the viscoelastic creep compliance at low stress has the form of Prony series,

$$\Delta S(\psi)=\sum_i C_i\left(1-\exp\left(-\frac{\psi}{\tau_i}\right)\right) \tag{3}$$

In equation (3) C_i are constants and τ_i are called retardation times.

For sufficiently small stresses, a linear viscoelastic region, where the VE-strains are proportional to the stress level, may exist. By normalizing the introduced constants and including the normalization constants in C_i it can be assumed that at very low stress $g_1 = a_\sigma = 1$. Regarding g_2 the assumption $g_2 = 1$ can be used at the lowest stress level investigated, which makes its value undetermined at lower stress values (including when stress is approaching to zero). Using these assumtion equation (1) becomes the strain-stress relationship for linear viscoelastic, nonlinear viscoplastic materials. The physical meaning of the damage related function $d(\sigma_{\max})$ in equation (1) was revealed in [22] analysing the elastic strain term.

The last term in equation (1) accounts for irreversible Viscoplastic (VP) strains that develop at high stresses and increase with the time of loading. In [21,23] the development of VP strains was successfully described by functions presented in [24]. In this model VP-strain grows during loading with specified time dependence and dependence on the applied stress given by:

$$\varepsilon_{VP}(\sigma,t)=C_{VP}\left\{\int_0^{t/t^*}\left(\frac{\sigma(\tau)}{\sigma^*}\right)^M d\tau\right\}^m \tag{4}$$

C_{VP}, M and m are constants to be determined, t/t^* is normalized time where t^* is an arbitrary chosen characteristic time constant and σ^* is an arbitrary chosen stress constant. In this study $t^* = 7200 s$ and $\sigma^* = 1$ MPa are assumed. It should be noted that equation (4) slightly differs from the original form: t^* and σ^* have been introduced to have C_{VP} dimensionless (if C_{VP} is presented in %, the calculated VP-strain values in equation (4) will be also in %).

When stress is kept constant, as in a creep test, $\sigma=\sigma_0$ and the integration in equation (4) is trivial. The VP-strain dependence of the creep test length, Δt_1 obeys the following power function:

$$\varepsilon_{VP}(t_1)=C_{VP}\left(\frac{\sigma_0}{\sigma^*}\right)^{Mm}\left(\frac{\Delta t_1}{t^*}\right)^m \tag{5}$$

If the material behaviour can be described by equation (4), the time dependence of VP-strains in constant stress creep test should follow a power function with respect to time as predicted by equation (5):

$$\varepsilon_{VP}(t_1) = A\left(\frac{\Delta t_1}{t^*}\right)^m \tag{6}$$

where A has power law dependence on the applied stress level in the creep test:

$$A = C_{VP}\left(\frac{\sigma}{\sigma^*}\right)^{Mm} \tag{7}$$

In order to obtain VP constants C_{VP}, M and m, multiple step creep and strain recovery tests must be performed. When the VP strain analysis is finished and model parameters determined, it is possible to subtract the VP-strain from the total strain, in order to obtain pure VE response. For VE analysis one step creep tests must be performed.

2.2. Experimental procedures
Even thou identification of parameters in the described model requires wide range of experiments, in this paper only experimental procedures referred in the "Discussion" section will be described.

2.2.1. Materials
The RCF produced by special type of the viscose process "Cordenka 700 Super 3" (Cordenka GmbH, Germany) are used as reinforcement in this work. Some of the characteristics of these fibres are available from manufacturer [25] as well as reported in [26-27].

Epoxidiezed pine oil based resin EpoBioX (Amroy, Finland) with Amroy Ca35Tg curing agent (mixing ratio 100:27) was used. This resin is approximately 75% bio-based. Resin plates were manufactured by use of resin transfer moulding. The resin was infused at room temperature and at low flow speed and then cured for 2h at 80°C.

Fibre preforms were manufactured by winding fibre roving on steel plate using filament winding machine (2 layers). Afterwards these preforms were impregnated using vacuum infusion with resin heated to 50°C. After impregnation plates were cured for 2h at 80°C. The composites had volume fraction of fibres $V_f = 67.1\%$ (average value of V_f is obtained from image analysis of micrographs).

2.2.2. Conditioning
Since bio-based materials are very sensitive to moisture, the influence of humidity was analysed. Non conditioned samples (NC) were tested as received (at room environment: relative humidity, RH≈24%, room temperature RT≈23°C). The conditioned samples were stored in an environment with controlled humidity until moisture content in materials reached equilibrium. Two RH levels, 41% and 70%, were used. Prior to conditioning, the specimens were kept in the oven at 50°C until they reached a constant mass. The fixed level of relative humidity was achieved by use of saturated solution of different salts. The weight of samples was regularly measured to ensure that moisture content reached saturation level.

2.2.3. Tensile tests
Quasi-static tensile tests of composites were performed in displacement controlled mode at 2mm/min (strain rate ≈2%/min) on electromechanical tensile machine Instron 3366 equipped with 10 kN load cell and pneumatic grips. Standard Instron extensometers 2620-601 (50mm base) were used to measure longitudinal strain. Fibre bundle tensile tests (gauge length of bundles 100 mm) were also performed on Instron 4411 in displacement controlled mode with loading rate 10mm/min (≈10%/min strain rate). The machine was equipped with mechanical grips and 500N load cell. Every bundle was fitted with end tabs - flat pieces of wood were glued at the each bundle end (Araldite 2011 two component epoxy adhesive was used). Since it was not possible to perform direct strain measurement

on bundles, the displacement of the cross-head of the tensile machine was used to calculate strain. In order to obtain actual strain value, machines compliance was measured and taken into account.

Tensile creep with following strain recovery tests for resins and composites were performed using creep rig with dead weight. Creep test on bundles were performed with the same set up as for quasi-static tensile tests. Two types of creep tests were performed: a) single step; b) multiple step. In the latter case a load is applied in steps. Each loading step was followed by recovery period, where no load is applied. Recovery interval was 8 times longer than loading interval. Multiple step creep test consisted of loading intervals with duration of 10, 20, 30 and 60 min. Single step creep tests were performed in one step with loading interval 120 min. The irreversible strains at the end of the recovery period (if present) were taken as VP-strain.

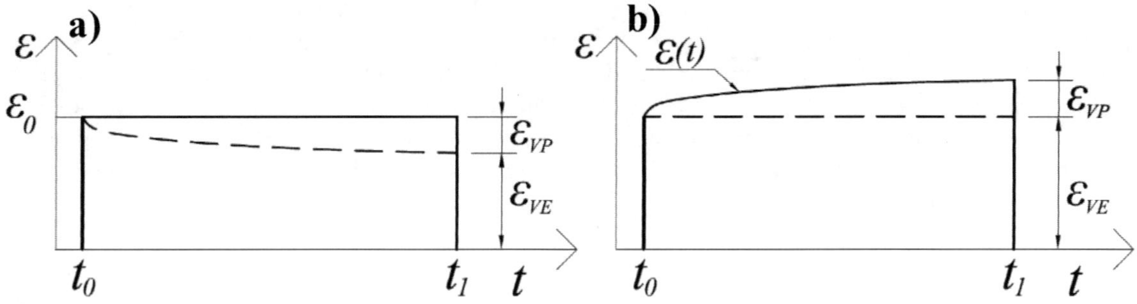

Figure 2. Schematic drawing of the relaxation test: a) with constant applied strain and b) strain as a function of time keeping constant viscoelastic strain.

3. Discussion

3.1. Complications with model formulation and parameter identification

The developed model is in the form of $\varepsilon(\sigma)$ (strain as a response to applied stress) and all the nonlinearity parameters in the model have to be determined as dependent on stress. Therefore all tests performed in order to obtain parameters for model must be stress controlled, which would be multiple-step and one-step creep tests at different load levels. The validation of the model also requires stress controlled tensile tests. These tests with different load rates are more difficult to perform than strain controlled experiments. It is also more common to perform strain/displacement controlled tensile tests. In addition most of codes for numerical structural analysis and also most of analytical micromechanics models (rule of mixtures, concentric cylinder assembly model) and even the classical laminate theory require constitutive model where stresses are expressed as a function of strains. The above arguments motivate development of a model in form $\sigma(\varepsilon)$. It is expected that due to the viscoelastic nature of the response the stress would be expressed through convolution integral with respect to the strain ramp and the relaxation function which is strain dependent. In order to obtain parameters for such a model, relaxation tests should be performed where viscoelastic (VE) strain should be kept constant. If material has no viscoplastic (VP) strains, then these tests are simple to perform – constant strain level in sample must maintain for certain time. However most of the materials have also VP strain component and performing relaxation test where VE strain is kept constant becomes very challenging. While the applied strain is maintained at the same level, as the time progresses, also VP strain will develop and VE strain will reduce with time (Figure 2a)). Therefore in order to obtain accurate results from relaxation tests, the overall applied strain level must be increased to compensate for development of VP strains, so that VE strain is constant during the test (Figure 2b)). Since the VP strain development depends on the stress-time dependence, see equation (4), which is not known before the test is finished, the VP strain time dependence can not be predicted even if the law for VP strains is identified in creep and strain relaxation tests. One way to access VP strain problem in the relaxation experiment is to perform this test several times. After first test some approximate information about VP strains will be available (the final value obtained experimentally after recovery and the stress-time dependence has been measured. The subsequent experiment should be modified by accounting for VP strain. However, this approach would only complicate the experimental program and increase the number of tests needed to obtain parameters. Fortunately in some cases VP strains can be neglected. For some materials it is possible to condition samples in a sequence of creep tests until amount of the new developed VP strain is negligible.

Assuming that the viscoelastic model in form of $\sigma(\varepsilon)$ is found, the application of it for any strain controlled ramp is straightforward: by integration of the time dependent viscoelastic strains over the required time interval. It is more difficult (impossible) to deal in this way with viscoplastic strain-stress because these strains are intrinsically caused by stresses and not the opposite. Viscoplastic strain value at certain instant of time depends on the stress history.

In a case when the applied strain has viscoelastic as well as viscoelastic part the best way is to use a discrete form of the material model using small strain increments to simulate the strain ramp. In this approach the stress in the time instant t_{k+1} is found knowing the strain and stress at t_k . This implies that the viscoplastic strain at t_k has been calculated using equation (4). The calculation is a multi-step procedure: a) at a given applied strain at t_{k+1} the viscoelastic strain is estimated as the difference between the applied and the viscoplastic strain at t_k; b) the viscoelasticity model is used to find stresses at t_{k+1} ; c) the found stress is used in the viscoplasticity law to find the viscoplastic strain at t_{k+1}. If the time step is small the accuracy is sufficient.

A general assumption that a region of linear viscoelasticity exists at low stresses and in the linear VE region functions $g_1 = g_2 = a_\sigma = 1$ has been used in many studies [19]. However, in thermodynamics there is neither requirement of the existence of linear region nor $g_1 = g_2 = a_\sigma = 1$. The requirement is that at very low stresses these functions have constant values and that the compliance there can be expressed in Prony series. Since the above constants are always multiplied by C_i in Prony series, for convenience one may include these constant values at zero in still unknown C_i and have at zero stress $g_1 = g_2 = a_\sigma = 1$. It should be noted that experimentally these conditions can be fulfilled for g_1 and a_σ but not for g_2. The reason is that the lowest stress level used in creep and strain recovery test for finding parameters cannot be zero. It has certain finite value,σ_1, to be able to perform reliable strain measurements and it is not known *a priori* if it belongs to a linear region or not. Inspecting the creep and strain recovery expressions for this first test, one can see that in expressions C_i is always multiplied by g_2 and therefore it is not possible to find both of them during the fitting procedure: one can find only the product. Certainly, one can define $g_2(\sigma_1) = 1$ and determine C_i values which are stress state independent. Using tests at different stress levels we find $g_2(\sigma)$ (which is actually $g_2(\sigma)/g_2(\sigma_1)$). Obviously, there is no way to find the g_2 values for $\sigma < \sigma_1$ because there are no test results. The only clear thing is that at zero stress there is no reason to expect $g_2 = 1$ (since in fact it is $g_2(0)/g_2(\sigma_1)$.

Another problem linked to these stress dependent functions in the material model is lack of physically based information regarding the shape of these functions. In different papers, different types of functions are presented [17-18] and the approach has been based on fitting accuracy rather than physical meaning.

3.2. Complications with material behaviour

Although RCF are continuous and have resemblance with filaments of other man-made fibres (glass, carbon etc.), mechanical behaviour of these fibres is completely different. Since RCF is of natural origin, they are very sensitive to surrounding environment. As can be seen in Figure 3a), where simple tensile tests for bundles at different relative humidity levels are presented, they are very sensitive to moisture. The stress-strain curves are about 20-30% lower at high RH. This behaviour of fibres reflects also in behaviour in composites (Figure 3b).

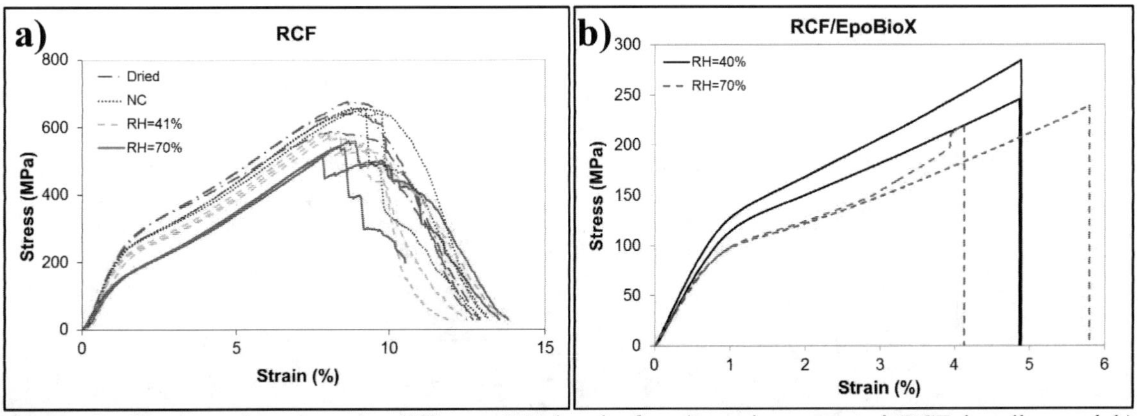

Figure 3. Stress-strain curves at different RH levels for a) nonimpregnated RCF bundles and b) RCF/EpoBioX composites.

Obviously, in order to develop model which is based on properties of constituents, moisture content in constituents and its effects must be studied. As described below the equilibrium moisture content in constituents, which is different in the fibre and in the matrix, depends on the relative humidity and also on the moisture diffusion related conditions at the fibre/matrix interface.

From the conditioning experiments moisture content at specific RH level was found and values are presented in table 1. Theoretically, moisture content in composite is related to the moisture in constituents by a simple rule-of-mixture (ROM) expression:

$$C_c^w = C_f^w W_f + C_m^w W_m \tag{8}$$

where C^w is moisture content and W is weight fraction. Indices c, m and f are for composite, matrix and fibre. Weight fraction of fibre can be calculated:

$$W_f = \frac{\rho_f}{\rho_c} V_f \tag{9}$$

Where ρ represents densities. Density of composite can be also found from ROM:

$$\rho_c = \rho_f V_f + \rho_m V_m \tag{10}$$

Density of EpoBioX resin $\rho_m = 1.06$ g/cm^3 and RCF fibres $\rho_f = 1.50$ g/cm^3. Values for moisture content in composite calculated according to equation (8) are also presented in table 1. It can be seen, that the composite has less moisture, than we obtained theoretically. This means, that at given RH fibres absorb less moisture, when they are in composite. Hence, in order to verify at certain RH the composite model, which is based on properties of constituents must be tested at correct environmental conditions. According to these results, fibres must be conditioned and tested at lower RH level, than composite and matrix. It is possible to back-calculate moisture content in fibres that are embedded in matrix by using equation (8): $C_w^f = 3.16\%$.

We can plot the RH-moisture content relationship for fibre bundles by using the two data points in table 1 and assuming, that at RH=0%, the moisture content in RCF is 0. As shown in Figure 4 we obtain almost linear relation between RH level and moisture content in RCF. From Figure 4 we can read that moisture content $C_w^f = 3.16\%$ corresponds to RH=21%, which is approximately RH in the room at the time of testing.

Table 1. Moisture content in material at different RH levels

Material	**RH=41%**	**RH=70%**
RCF	6.41%	10.40%
EpoBioX	0.28%	0.82%
RCF/EpoBioX (experimental)	2.41%	5.91%
RCF/EpoBioX (theoretical)	4.82%	7.91%

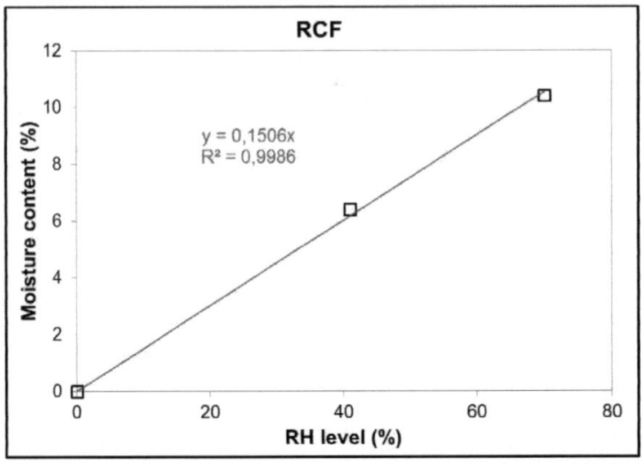

Figure 4. Moisture content in RCF as a function of RH level.

RCF also exhibit highly non-linear time-dependent behaviour. It is clearly visible in Figure 5a), where curves from loading-unloading tests are presented. It is also obvious that the slope of the stress-strain curve from loading and unloading differs a lot. None of these slopes represent the elastic modulus of the material, but the unloading slope even in the low strain region is more affected by the viscoelastic behaviour. For the composite the hysteresis loop is much smaller than for RCF bundles; however there is noticeable difference in loading/unloading slopes of stress-strain curve, as shown in Figure 5b).

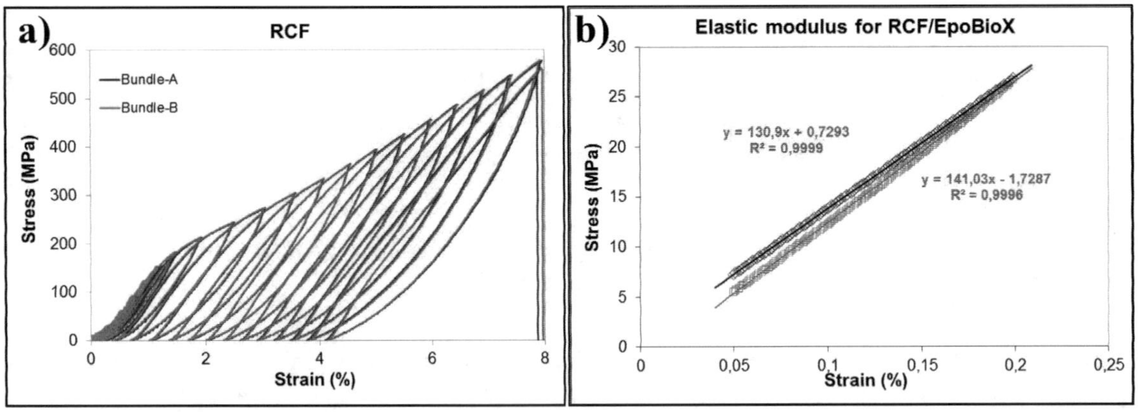

Figure 5. Stress-strain curves for a) loading-unloading test for fibre bundle and b) composite stress-strain response in small stress region

4. Conclusions

Experimental and theoretical challenges related to development of material model for regenerated cellulose fibre (RCF) reinforced composites with bio-matrix are identified and potential solutions are suggested.

The theoretical model suggested for description of the composite time dependent behaviour is based on multiscale analysis: Schapery's type nonlinear viscoelastic models generalized for viscoplastic strains and damage are developed separately for fibre and for matrix characterization and used to find the response of the composite which is also nonlinear viscoelastic and viscoplastic with damage.

In order to use this approach the material models for constituents developed in formulation where the stress is the independent variable have to be inverted to form where stress is calculated as a function of the strain ramp.

Two directions towards this goal are identified and critically analysed: a) using series of relaxation tests at different applied strain levels instead of creep test; b) incremental formulation of the earlier developed models based on creep tests with following inversion to calculate stress versus the applied strain.

The fibre behaviour is very complex with large viscoplastic and viscoelastic strains and hysteresis loops developing in high stress cycling. The model development is further complicated by the high moisture sensitivity of the RCF composite constituents. It is demonstrated that the fiber stress-strain response at different moisture contents has to be analysed before the composite model can be applied for a certain relative humidity.

Acknowledgments

Part of this study was financially supported by Interreg IVA Nord project ANACOMPO funded by EU and local government. Runar Långström and Peter Mannberg (Swerea SICOMP, Pitea, SWEDEN) are acknowledged for manufacturing of resin and composite plates, Newsha Doroudgarian is acknowledged for her help with experiments.

References

[1] Andersons J, Sparnins E, Joffe R and Wallstrom L 2005 Strength distribution of elementary flax fibres *Compos Sci Technol* **65** 693-702

[2] Summerscales J, Dissanayke N P J, Virk A S and Hall W 2010 A review of bast fibres and their composites. Part 1 – Fibres as reinforcement, *Compos: Part A-APPL S* **41** 1329-1335.

[3] Bos H L and Van Den Oever J A, Peters O C J J 2002 Tensile and compressive properties of flax fibres for natural fibre reinforced composites *J Mater Sci* **37** 1683-1692.

[4] Magurno A 1999 Vegetable fibres in automotive interior components *Die Angewandte Makromolekulare Chemie* **272** 99–107.

[5] Malnati P 2009 ECO Elise concept: lean, speedy and green *Composites technology* **15**.

[6] Marsh G 2003 Next step for automotive materials *Materials Today* **6** 36-43.

[7] Hepwort D G, Bruce D M, Vincent J F V and Jeromidis G 2000 Manufacture and mechanical testing of thermosetting natural fibre composites *J Mater Sci* **35** 293-298.

[8] Oksman K 2001 High quality flax fibre composites manufactured by the resin transfer moulding process *J Reinf Plast Comp* **20** 621-627.

[9] Fombuena V, Sánchez-Nácher L, Md S, Juárez D and Balart R 2012 Study of the Properties of Thermoset Materials Derived from Epoxidized Soybean Oil and Protein Fillers *J Am Oil Chem Soc* 1-9 (article in press).

[10] Raquez J M, Deleglise M, Lacrampe M F and Krawczak P 2010 Thermosetting (bio)materials derived from renewable resources: a critical review *Prog Polym Sci* **35** 487-509.

[11] Joffe R, Nyström B, Erwan Le Marec P Investigation of damage accumulation in flax fibre composite laminates in *20th Conference on Manufacturing and Design of Composites* (Piteå, Sweden, June 8-9, 2009).

[12] Nyström B, Joffe R, Rozite L, Gong G Development of bio-based advanced composites with transport structures in mind in *The 22nd Annual International SICOMP Conference* (Göteborg, Sweden, May 30-31, 2011).
[13] Charlet K, Baley C, Morvan C, Jernot J P, Gomina M and Breard J 2007 Characterization of Hermes flax fibres as a function of their loacation in the flax stem and properties of the derived unidirectional composites *Compos: Part A-APPL S* **38** 1912-1921.
[14] Charlet K, Jernot J P, Gomina M, Bréard J, Morvan C nad Baley C 2009 Influence of an Agatha flax fibre location in a stem on its mechanical, chemical and morphological properties *Compos Sci Technol* **69** 1399–403.
[15] Sparnins E and Andersons J 2009 Diameter variability and strength distribution of elementary flax fibers *J Mater Sci* **44** 5697–5699.
[16] Lou Y C and Schapery R A 1971 Viscoelastic Characterization of a Nonlinear Fiber-reinforced Plastic *J Compos Mater* **5** 208-234.
[17] Schapery R A 1997 Nonlinear Viscoelastic and Viscoplastic Constitutive Equations Based on Thermodynamics *Mechanics of Time-Dependent Materials* **1** 209-240.
[18] Varna J, Rozite L, Joffe R and Pupurs A 2012 Nonlinear behavior of PLA based flax composites *Plast. Rubber Compos* **41** 49-60.
[19] Varna J, Sparnins E, Joffe R, Nattinen K and Lampinen J 2012 Time dependant behavior of flax/starch composites *Mechanics of time-dependent materials* **16** 47-70.
[20] Giannadakis K, Mannberg P, Joffe R and Varna J 2011 The source of inelastic behavior of Glass Fibre/Vinilester non-crimp farbic [±45]s laminates *J Reinf Plast Comp* **30** 1015-1028.
[21] Marklund E, Eitzenberger J and Varna J 2008 Nonlinear viscoelastic viscoplastic material model including stiffness degradation for hemp/lignin composites *Compos Sci Technol* **68** 2156-2162.
[22] Varna J 2011 Characterization of vicoelasticity, viscoplasticity and damage in composites in *Creep and Fatigue in Polymer Matrix Composites* ed Gueded R M (UK: Woodhead Publishing Materials) pp 514-542.
[23] Nordin L O and Varna J 2006 Nonlinear viscoplastic and nonlinear viscoelastic material model for paper fiber composites in compression *Compos: Part A-APPL S* **37** 344-355.
[24] Zapas L J and Crissman J M 1984 Creep and Recovery Behavior of Ultra-high Molecular Weight Polyethylene in the Region of Small Uniaxial Deformations *Polymer* **25** 57-62.
[25] CORDENKA® High Tenacity Rayon Filament Yarn data sheet from CORDENKA GmbH.
[26] Gindl W, Reifferscheid M, Adusumalli R B, Weber H, Röder T, Sixta H and Schöberl T 2008 Anisotropy of the modulus of elasticity in regenerated cellulose fibres related to molecular orientation *Polymer* **49** 792-799.
[27] Joffe R, Andersons J and Sparnins E Applicability of Weibull strength distribution for cellulose fibers with highly non-linear behaviour In *Proc. of the 17th International Conference on Composite Materials (ICCM-17)* (Edinburg, UK, 27-31 July 2009).

IOP Conf. Series: Materials Science and Engineering **48** (2013) 012008 doi:10.1088/1757-899X/48/1/012008

Correlation between mechanical and dielectric properties of Alfa/Wool/Polymeric hybrid fibres reinforced polyester composites

A TRIKI[1], M A OMRI[1], M GUICHA[2], M BEN HASSEN[2], M AROUS[1] and A KALLEL[1]

[1]Laboratoire des matériaux composites, polymères et céramiques, FSS 3018, Tunisia

[2]Laboratoire de génie textile, Université de Monastir, ISET Ksar Hellal, Tunisia

E-mail: trikilamacop@yahoo.fr

Abstract Dielectric measurements and tensile testing of polyester/natural fibres (Alfa/wool) and thermo binder fibres (Pe/Pet) composites were investigated in order to study the adhesion of the fibres in the polyester matrix. Two composites #1 and #2 having 17:1:2 and 17:2:1 as a relative fraction of alfa/wool and thermo binder (Pe/Pet), respectively, have been characterized in this study. The obtained results revealed that the fibres adhesion in the matrix was better in the composite #1 than in the composite #2. Indeed, the analysis of the interfacial or Maxwell-Wagner-Sillars (MWS) polarization intensity, using the Havriliak–Negami model, has shown a lower intensity and the tensile testing exhibited a higher Young modulus in the composite #1. So the thermo binder fibres improve this adhesion.

1. Introduction

Over the last two decades, natural fibres reinforced polymer composites have been the focus of academic and industrial research interest (e.g. automotive, sport, construction and packaging) due to several advantages, such as low cost, high strength-to-weight ratios, low densities and recyclability compared to the synthetic fibre composites. Mechanical properties of fibre reinforced composites depend mainly upon the nature of fibre, matrix and fibre/matrix adhesion. Natural fibres such as alfa, hemp, sisal, flax, kenaf and jute are highly hydrophilic owing to the presence of hydroxyl groups (OH). This character is a potential cause for poor interfacial adhesion with hydrophobic polymer matrices. This makes natural fibre reinforced polymer composites vulnerable to environmental attacks, which may cause weakness and thus reduce its life span. In other words, insufficient adhesion between the hydrophobic polymers and hydrophilic fibres results in poor mechanical properties in these composites. Interfacial adhesion and resistance to moisture absorption of natural fibre composites can be improved by the suitable modification of fibres surface, or by using coupling agent.

In addition, the amount of moisture sorption can be reduced significantly through the replacement of natural fibres with a small amount of synthetic fibres, such as glass or carbon. Many techniques have been used to give evidence for the effect of these treatments on the fibres/matrix interfacial adhesion. The most frequently ones used are dynamic mechanical analysis, scanning electron microscope (SEM) observation, dielectric spectroscopy and tensile testing analysis.

Composites having two or more fillers contained in the same matrix are called hybrid composites. Recently there has been a growing interest in hybridizing different natural fibres in order to produce high performance composite materials. Hybrid biocomposites can be designed by the combination of a synthetic fibre and natural fibre (biofibre) in a matrix, or a combination of two natural fibre / biofibre in a matrix. Hybridization with glass fibre provides a method to improve the mechanical properties of natural fibre composites and the effect on different modes of stress depends on the design and construction of the composites [1]. For example, the hybridization of oil palm fibres with glass fibres increases the tensile strength, Young's modulus, and elongation at break of the hybrid composites [2]. Moreover, synergistic strengthening of fibres has been observed in the case of banana/sisal hybrid fibre composites, where tensile and flexural properties show a positive hybrid effect [3, 4].

This research work aims at investigating the effect of wool and thermo binder (Pe/Pet) fibres on the fibres/matrix interfacial adhesion of the composites of polyester/natural fibres, where the natural fibres consist of a mixture of Alfa and wool, and the polymeric fibres composed of PET and PE as a cover. To do this the study compares two composites differing in the relative volume fractions of the wool and thermo binder fibres. These composites were characterized by the dielectric spectroscopy and the tensile testing analysis. The dielectric properties of these composites were analyzed and the intensity of the interfacial dielectric relaxation was determined, using the Havriliak-Negami model, in order to examine the interfacial region between the fibres and the matrix. A correlative study between dielectric and tensile properties of these composites was realized to confirm this dielectric study.

2. Materials and experimental techniques

2.1. Materials

The unsaturated polyester matrix used in this research work was supplied by Cray Valley/Total and is the same as that used in our previous study [5]. The matrix was mixed with the initiators methyl ethyl ketone peroxide and cobalt octanone at a concentration of 1.5% w/w before the introduction of the alfa fibres. The latter which were extracted from the plant were treated chemically by an NaOH solution and bleached in an NaClO solution. Then, they were separated mechanically with a Shirley analyzer. As combining nonwoven fibres with only alfa fibres was not possible because of the non-cohesion between them, the alfa fibres were mixed with wool fibres to ensure cohesion. Moreover, since natural fibres on their own cannot be thermoformed and require the addition of polymeric fibres to act as binders, the added thermo binder fibres were composed of PET and PE as a cover. The obtained sheet of nonwoven fibres was made up of the three kinds of fibres under study. The diameters of the alfa and wool fibres were 204.86 and 37.21 μm, respectively. Besides, the length of the thermo binder fibres was 51 mm; the fibres' count was 4 deniers, and their melting temperatures were 260°C and 110°C. The relative volume fractions of these fibres in dry preforms used in the composites #1 and #2 had a ratio of alfa to wool to PET–PE of 17:1:2 and 17:2:1, respectively. Regarding the preparation steps of the nonwoven fibres (alfa + wool + PET-PE) fabric, they were given in details in our previous work [6]. The obtained composites #1 and #2 manufactured with the classical contact mould again as previous described [6], had 4.8 and 4.9 wt % fibre fractions, respectively.

2.2. Experimental techniques

Dielectric measurements were carried out using an Alpha dielectric–impedance analyzer (Novocontrol), with the measurements of the studied samples taken over the temperature range from the ambient to 150°C and in a frequency interval from 10^{-1} to 10^{6} Hz. A circular gold electrode (2 cm in diameter) was sputtered on both surfaces of the sample to ensure good electrical contact with the gold-plated measuring electrodes. A sinusoidal voltage was applied in the sample in order to create an alternating electric field that produced polarization which oscillated at the same frequency as the electric field but had a phase angle shift (δ). Such phase angle shift was measured by the comparison of the applied voltage with the measured current, which was divided into capacitive and conductive components. The dielectric parameters were calculated with the following equations:

$$\varepsilon^* = \varepsilon' - j\,\varepsilon'' \tag{1}$$

$$\varepsilon' = \frac{C_p(sample)d}{\varepsilon_0 A} \tag{2}$$

$$\varepsilon'' = \frac{G_{(Sample)}\, d}{\omega\, \varepsilon_0\, A} \tag{3}$$

where ε′ and ε″ are the real and imaginary parts of the complex permittivity ε*. Tan δ = ε″/ε′ is the dissipation factor. A and d are the area and thickness of the sample, respectively. C_p is the capacitance, G is the conductance, $\varepsilon_0 = 8.854\times10^{-12}$ F m^{-1} is the permittivity of the free space.

The dielectric experiments were carried out with fixed temperatures and scanning frequencies from 10^{-1} to 10^{6} Hz. The tensile testing of the composites #1 and #2, it was carried out with a Lloyds Dynamometer universal testing machine as per NF T 57-301 at a crosshead speed of 5 mm/min and a gripping length of 100 mm [7]. The specimen was cut out in the direction of nonwoven production. All of the results were calculated at the average of 10 samples for each test.

3. Results and discussion

3.1. Dielectric properties

Comparative plots of the frequency dependence of the dielectric permittivity ε' and those of the dissipation factor, tan δ, in the composites #1 and # 2 for different temperatures varying from 40°C to 150°C in increments of 10°C are depicted in Figure 1-(a-d), respectively. For low frequencies, the dielectric permittivity ε' reaches high values with the increase in temperatures. It is worthy to note that at low temperatures, ε' is steadily constant and approximately equal to that of the polyester resin in the composite #1. However, in the case of the composite #2, the dielectric permittivity ε' is superior to that of the resin. This indicates the presence of water dipoles polarization in the composite #2. Indeed, the dissipation factor, tan δ, proves the presence of a dielectric relaxation at low temperatures as illustrated in Figure 1-d. As temperature increases, the dissipation factor curves of the composite #1 have revealed the presence of a dielectric relaxation attributed to interfacial polarisation field effect-Maxwell- Wagner- Sillars (MWS) polarization effect added to the one associated with dc conductivity effect above glass transition temperature (T_g). The interfacial dielectric relaxation was the result of the charge accumulation between the fibres and the matrix having different conductivities and permittivities. Therefore, the enhancement of these relaxations above T_g temperature amplifies ε' intensity as temperature increases. In the case of the composite #2, the isothermal runs of the dissipation factor show the presence of only one dielectric relaxation at high temperatures associated with the MWS polarization effect. This results in a lower rise of the permittivity ε' at low frequencies and high temperatures in comparison to that of the composite #1.

The incorporation of natural fibres in the matrix has also led to a decrease in the intensity of the dissipation factor in both composites compared to that of the matrix [6]. This decrease is more important in the composite #1, which can be explained by a better thermal insulation in this composite. Indeed, it is recognized that the measurement of dissipation factor of insulating material is important since the loss tangent is a measure of the alternating current electrical energy which is converted to

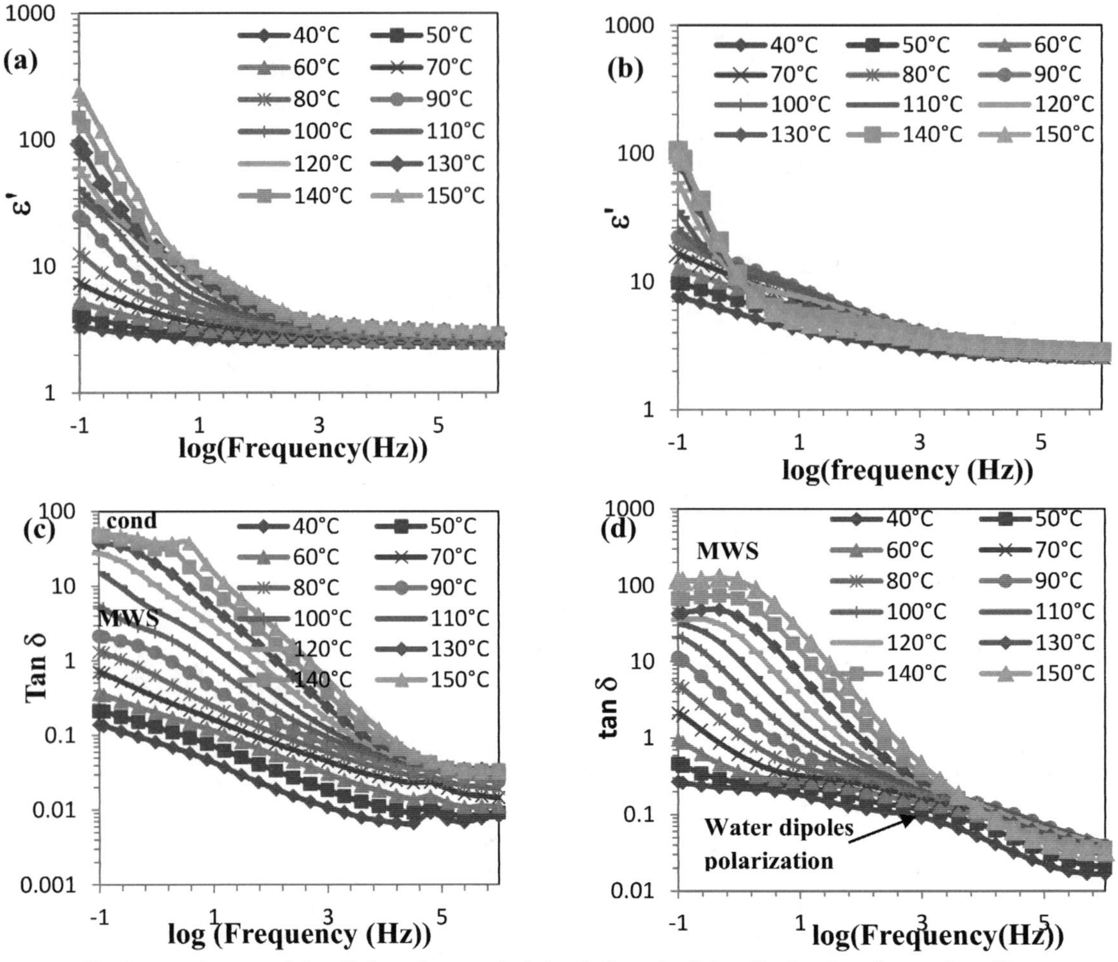

Figure 1 Isothermal runs of the dielectric permittivity (ε') and of the dissipation factor (tan δ) versus frequency for the composite #1 (a), (c) and composite #2 (b) and (d).

heat in an insulator. Such heat raises the insulator temperature and accelerates its deterioration.

To minimize the effects of the dc conductivity, the formalism of ''electric modulus'' or ''inverse complex permittivity'' is introduced. This electric modulus has recently been adapted for the investigation of dielectric processes occurring in composite polymeric systems, as well as proposed for the description of systems with ionic conductivity. The electric modulus, M^*, is defined by the following Eq. 4:

$$M^* = \frac{1}{\varepsilon^*} = \frac{1}{\varepsilon' - j\,\varepsilon''} = \frac{\varepsilon'}{\varepsilon'^2 + \varepsilon''^2} + j\,\frac{\varepsilon''}{\varepsilon'^2 + \varepsilon''^2} = M' + j\,M'' \quad (4)$$

where M' and M'' are the real and imaginary parts of the electric modulus, respectively. An advantage of using the electric modulus to interpret bulk relaxation properties is that the variation in the large values of the real part of permittivity and the loss factor at low frequencies are minimised. In this way, common difficulties of electrode nature and contact, space charge injection phenomena and absorbed impurity conduction effects, which appear to hide relaxation in the permittivity representation, can be solved or even ignored [8].

Adopting this electric modulus formalism, Figure 2-(a, b) shows the behaviour of M'' as a function of frequency for the composites #1 and #2 when they were heated over the temperature range from

40 °C up to 150 °C. A series of two distinct relaxations can be considered in the composite #1, the first of which is related to the α relaxation associated with the glass rubbery transition of the polymer and the second one, which appears at high temperatures, is attributed to ionic conduction effect. Yet, in the composite #2, a series of three distinct relaxations can be identified. Indeed, the α relaxation of the polymer appeared at temperatures above the glass transition temperature. The one which appeared at low temperatures is associated with the water dipoles polarization and the last one which appeared at high temperatures was assigned to the interfacial polarization effect. To further support these assignments, the activation energy relative to the different relaxation was evaluated using the Arrhenius relation:

$$f_{max} = f_o \exp\left(-\frac{E_a}{k_B T}\right) \tag{5}$$

where k_B, T, E_a and f_o are Boltzmann constant, temperature, activation energy of the process and frequency at high temperature. Figure 3 shows the evolution of log (f_{max}) versus 1/T for each one of the different observed relaxations; i.e., α and conductivity for the composite #1 and α and interfacial polarization effect for the composite #2. The mean values of the activation energy E_a and the relaxation time τ_o relative to α relaxation are 136.02 KJ mol^{-1}, $10^{-20.96}$s and 132.694 KJ mol^{-1}, $10^{-20.73}$s for the composites #1 and #2, respectively as mentioned in Table 1. Nevertheless, it can be noted that

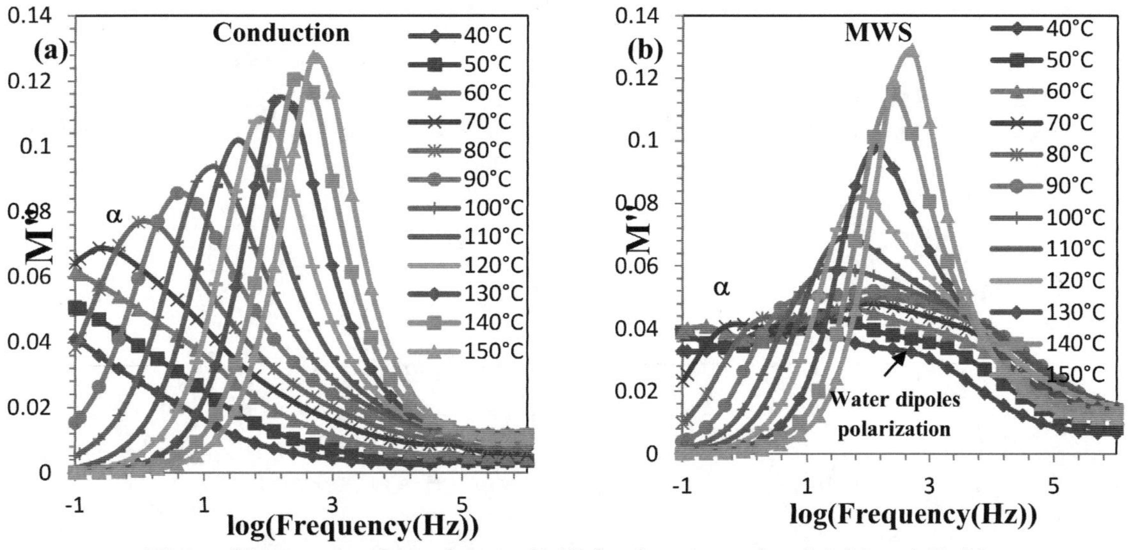

Figure 2 Frequency dependence of M'' for the composites # 1 (a) and #2 (b).

the incorporation of fibres decreases the apparent activation energy E_a^{α} for the two composites in comparison with that for the matrix determined in the previous study and this could be ascribed to the interaction between the fibres and matrix [5]. This interaction will determine the nature of the interfacial adhesion region as well as the interfacial polarization effect for the composite #2 i.e., 82.711 KJ mol^{-1} for E_a and $10^{-13.65}$s for τ_o. These values seem to be a typical thermoactivated phenomenon that is not related to an elementary relaxation of dipoles as the characteristic time τ_o is slightly different from that of Debye (10^{-13} s). In the case of composite #1, the average values of activation energy and time relaxation relative to conduction effect and are 96.76 kJ/mol and $10^{-15.52}$s, respectively. These values are in agreement with those reported in other research works [9].

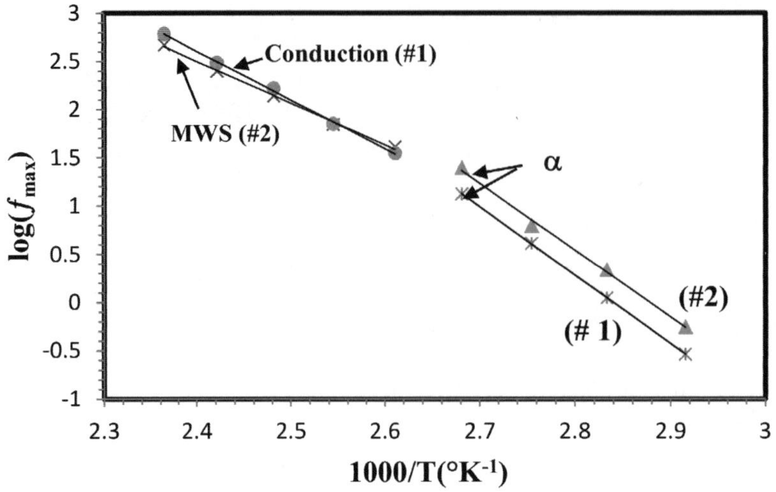

Figure 3 Arrhenius plots of the frequency of M'' versus the reciprocal temperature for the composites #1 (α relaxation (*), dc conduction (•)) and #2 (α relaxation (▲, MWS polarisation effect (×))

The Argand representation is used to analyse the nature of relaxation at high temperatures. Cole–Cole plots of the composites #1 and #2 at 150 °C are depicted in Figure 4-(a, b). It is well established that the response of every relaxation mechanism can be represented very precisely by a model function with four parameters at the most. Among others, the function:

$$\varepsilon^* = \varepsilon_\infty + \frac{\varepsilon_S - \varepsilon_\infty}{(1 + (i\omega\tau)^\alpha)^\beta} \tag{6}$$

Table 1 Activation Energies E_a and Relaxation Times τ_o for the Composites # 1 and #2.

Samples	Relaxation	E_a(KJ mol^{-1}	τ_o(s)
Composite#1 :			
	α relaxation	136.02	$10^{-20.96}$
	Conduction	96.33	$10^{-15.47}$
Composite#2 :			
	α relaxation	132.694	$10^{-20.73}$
	MWS	82.711	$10^{-13.65}$

introduced by Havriliak–Negami is widely used thanks to its suitability for mathematical processing. ε_S and ε_∞ are the dielectric constants on the low and high frequency sides of the relaxation, τ is the central relaxation time, ω is the radial frequency, α and β are fractional shape parameters describing the skewing and broadening of the dielectric function, respectively. Both α and β range between 0 and 1. These coefficients act for the deviation from the Debye equation. In fact, when α and β = 1, this equation reduces to the Debye equation. In the electric modulus formalism, the Havriliak–Negami equations (7, 8) have the following form:

$$M' = M_\infty \frac{[M_s A^\beta + (M_\infty - M_s)\cos\beta\varphi]\, A^\beta}{M_s^2 A^{2\beta} (M_\infty - M_s) M_s \cos\beta\varphi + (M_\infty - M_s)^2} \tag{7}$$

$$M'' = M_\infty M_s \frac{[(M_\infty - M_s)\sin\beta\varphi]\, A^\beta}{M_s^2 A^{2\beta} (M_\infty - M_s) M_s \cos\beta\varphi + (M_\infty - M_s)^2} \tag{8}$$

Where

$$M_s = \frac{1}{\varepsilon_s} \tag{9}$$

$$M_\infty = \frac{1}{\varepsilon_\infty} \tag{10}$$

$$A = [1 + 2(\omega\tau)^{1-\alpha} \sin\frac{\alpha\pi}{2} + (\omega\tau)^{2(1-\alpha)}]^{\frac{1}{2}} \tag{11}$$

$$\varphi = arctg[\frac{(\omega\tau)^{1-\alpha}\cos\frac{\alpha\pi}{2}}{1 + (\omega\tau)^{1-\alpha} sin\frac{\alpha\pi}{2}}] \tag{12}$$

Accordingly, the dotted curves are produced by the best fitting experimental points using the Havriliak–Negami equations (7, 8). In Figure 4-(a, b), it is illustrated that it is impossible to fit the Havriliak–Negami model to all the experimental points for these composites. So, two semicircles are obtained in every examined temperature. This analysis confirms the presence of the MWS relaxation which is overlapped by dc conductivity effect in composite #1. The parameters evaluated by fitting data are listed in Table 2. To determine the parameters characteristics of the Havriliak and Negami model (α, β, M_s, M_∞), the experimental M' and M'' data are smoothed through a numerical simulation in the complex plane. The purpose of such simulation is to find the theoretical values (M'_{th}, M''_{th}). Concerning the values of α, β, M_s and M_∞ which smoothed best the Havriliak–Negami data, they are

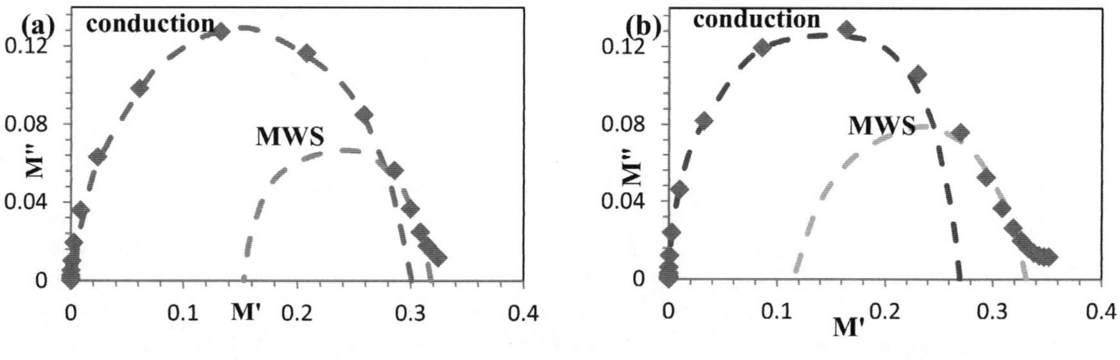

Figure 4 Argand's plots of M* of the composites #1(a) and #2 (b) at 150°C.

obtained by a successive approach method where the following expressions (13, 14) are minimized:

$$\chi^2_{M'} = \sum_i (M'_{th} - M'_{exp})^2 \tag{13}$$

$$\chi^2_{M''} = \sum_i (M''_{th} - M''_{exp})^2 \tag{14}$$

It has been proven that only one-quadruplet value is able to tone with these conditions. While the values of α and β obtained for the conductive effect fit with a pure Debye-type, the values of α and β obtained for interfacial polarization are in accordance with the Havriliak–Negami response.

IOP Conf. Series: Materials Science and Engineering **48** (2013) 012008 doi:10.1088/1757-899X/48/1/012008

Table 2. Parameters evaluated by data fitting according to the Havriliak–Negami equation for the composites #1 and #2.

Samples	T(°C)	relaxation	α	β	M_s	M_∞
Composite #1	130	cond	0.9915	0.90	0.002	0.295
		MWS	0.901	0.88	0.185	0.3191
	140	cond	0.995	0.93	0.002	0.298
		MWS	0.999	0.92	0.21	0.316
	150	cond	0.999	0.904	0.0033	0.301
		MWS	0.99	0.863	0.153	0.3182
Composite #2	130	cond	0.995	0.83	0.005	0.32
		MWS	0.92	0.892	0.199	0.315
	140	cond	0.999	0.819	0.00399	0.3269
		MWS	0.92	0.892	0.199	0.315
	150	cond	0.945	0.879	0.003	0.33
		MWS	0.92	0.892	0.199	0.315

In order to probe the interface between the matrix and the fibres, we have calculated the strength of the MWS relaxation in permittivity values defined by ($\Delta\varepsilon = \varepsilon_s - \varepsilon_\infty$) [10]. The obtained values for both composites are presented in Table 3. We can notice that the relaxation strength of MWS relaxation increases with the increase in temperature in both composites. This can be ascribed to the increase of free charges that come to be blocked in greater numbers at the interfaces, thus increasing the aptitude of the dipoles to be polarized. Furthermore, the strength of this relaxation is lower in the composite #1 compared to that of the composite #2 at each temperature. A similar trend is seen after a chemical treatment of alfa fibres indicating a regidification of fibre-matrix interfacial region which in turn reduce the ability of dipole to relax [11] and then enhance the compatibility and the adhesion between epoxy matrix and alfa fibres. As in the composite #1, the relative volume fraction of the Pe-Pet thermo binder fibres is twice the one of wool fibres contrary to the composite #2 in which the relative volume fraction of wool fibres is twice the one of thermo binder Pe-Pet fibres. The latter can enhance the adhesion of the fibres in the matrix. Contrary to the wool fibres which can reduce this interaction and this can be attributed to its hydrophilic character. Indeed, while the dielectric study has revealed the presence of the polarization of the water dipoles at low temperatures in the composite #2, this dielectric relaxation is absent in the composite #1.

Table 3. Interfacial relaxation strength of the relaxation MWS

Samples	T(°C)	Δε
Composite #1 :	120	1.85
	130	2.27
	140	1.59
	150	3.39
Composite #2 :	120	2.69
	130	4.11
	140	6.00
	150	5.67

3.2. Tensile properties

The interface between reinforcing fibre and the matrix plays a critical role in the mechanical performance of composites. A weak interface generally results in low strength and stiffness but high resistance to fracture, whereas a strong interface produces high strength and stiffness, but often low resistance to fracture. Figure 5-(a-c) shows the Young's modulus, strength, and stress at break of the composites #1 and #2, respectively. From these histograms, we can see that the Young's modulus of the composite #1 is superior to that of the composite #2. This shows a good cohesion of the materials to transfer stress from the matrix to the fibres [12]. However, the strength and stress at break of the composite #2 is superior to that of the composite #1. In a hybrid composite, the properties of the composite are mainly dependent on the modulus and the percentage elongation at break of the individual reinforcing fibres. The modulus of the Pe-Pet thermo binder fibre is comparatively higher than that of the alfa and the wool fibres [13], whereas the extensiveness of alfa fibre is low compared to the Pe-Pet thermo binder fibre and the wool fibres [13]. This results in early Alfa fibre failure which transfers stress to the stronger thermo binder and wool fibres. As the Pe-Pet thermo binder fibre is less extensive than that of wool fibre, it leads to its failure and it transfers high stress to the less strong wool fibres. This leads to the failure of the wool fibres resulting in the failure of the composite. The

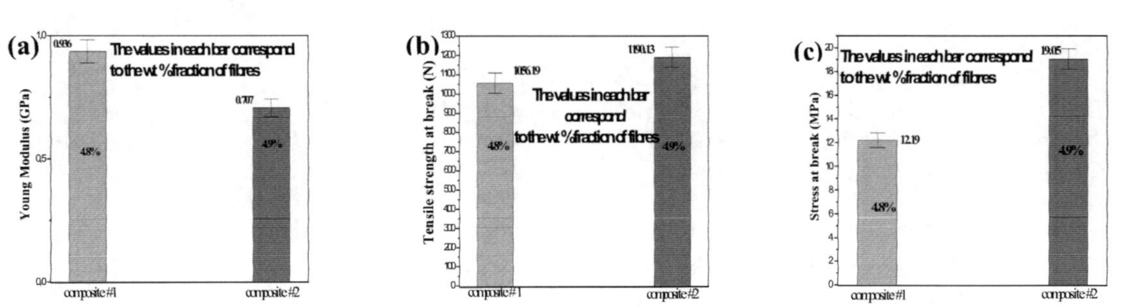

Figure 5 (a) Young's modulus, (b) tensile strength at break, and (c) stress at break of the composites #1 and #2.

enhancement of the tensile strength of the composite #2, i.e. at low volume fraction of the Pe-Pet thermo binder fibre loadings, will be due to the fact that at this particular composition, the wool fibre can effectively transfer the load from the Pe-Pet thermo binder fibre. Hence, comparatively less volume faction of Pe-Pet thermo binder fibres along with the wool fibres reinforcement give rise to a composite with high tensile strength, which in turn can result in an enhancement of the stress at break. A similar trend was obtained in the tensile strength of the sisal/glass hybrid polyester composite [14]. Accordingly Misha et al. have proven that the improvement in tensile strengths of the composites at low wt.% of glass fibre loadings will be due to the fact that at these particular compositions, the sisal fibre can effectively transfer the load from the glass fibre. The determination of the percentage elongation at break of the composites #1 and #2, as mentioned in Table 4, reveals that the elongation of the composite #1 is inferior to that of the composite #2 and this can be explained by a strong interface fibres/matrix [15, 16]. Indeed, a better adhesion of fibres in the matrix prevents sliding effect of the reinforcement in the matrix under the effect of stress. The tensile performance of the composite #1 becomes more marked than that of the composite #2 when specific properties are compared, as illustrated in Table 4. The difference between the specific stresses of both composites is lowered whereas that of the specific Young's modulus is increased in favour of the composite #1. So the adhesion of the reinforcement in the matrix is improved by the increase in the volume fraction of Pe-Pet thermo binder fibres.

Table 4. Tensile Properties of the Composites #1 and #2.

Samples	ε (%)	σ_t(MPa)	σ_t/ρ (MPa cm^3/g)	E_t(GPa)	E_t/ρ (GPa cm^3/g)
#1	4.70±0.23	12.19±0.61	12.0±0.6	0.936±0.046	0.92±0.05
#2	11.15±0.55	19.05±0.95	15.8±0.8	0.707±0.035	0.58±0.03

In table 4, σ_t, ρ and E_t are the stress at break, the density and the Young's modulus of the composites, respectively

4. Conclusion

The dielectric study of the composite #1 substantiates the presence of three dielectric relaxations, which were attributed to the α relaxation of the polymer, the ionic conduction which occurred above the glass transition and at low frequencies and the interfacial polarization effect, respectively. The dielectric study of the composite #2 have revealed, in addition, the presence of water dipoles polarisation which could diminish the adhesion of the reinforcement in the matrix as demonstrated by the interfacial polarisation analysis. The correlation of these dielectric properties to the tensile ones has confirmed that the increase of relative volume fraction of Pe-Pet thermo binder fibres to that of wool fibres could enhance the adhesion of the reinforcement in the polyester matrix.

5. Reference

[1] Singh B, Gupta M, Verma A 1995 *Const. Build. Mater.* **9** 39-44
[2] Bakar A, Hariharan H P S, Khalil A 2005 *J. Comp. Mater.* **39** 8 663-84
[3] Idicula M, Neelakantan N R, Thomas S In: *Proceedings of the USM JIRCAS* joint international symposium; 2001. p. 368.
[4] Idicula M, Malhothra S K, Joseph K, Thomas S 2005 *Comp Sci Technol* **65** 1077-84
[5] Triki A, Guicha M, Ben Hassen, M, Arous M, Fakhfakh Z 2011 *J. Mater. Sci* **46** 3698-3707
[6] Triki A, Guicha M, Ben Hassen M, Arous M 2012 *J. Appl. Poly. Sci* DOI: 10.1002/app.38499
[7] Guicha M, Hlimi M T, Ben Hassan M, Sakli F. (to INNORPI.). Pat. Appl. TN 2009/0200 (2000; in accordance with law number 2000–84).
[8] Tsangaris G M; Psarras G C, Kouloubi N 1998 *J. Mater. Sci.* **33** 2027-37
[9] Ben Amor I; Rekik H; Kaddami H; Raihane M, Arous M, Kallel A 2009 *J. Electrostat.* **67** 717-22
[10] Puertolas J A, Castro M, Telleria I, Algeria A 1999 *J. Polym. Sci.* B *37* 1337-49
[11] Arous M, Ben Amor I, Boufi S, Kallel A 2007 *J. Appl. Polym. Sci.* **106** 3631-41

[12] Arbelaiz A, Fernandez B, Cantero,G, Llano-Ponte R, Valea, A, Mondragon I *Compos. A* 2005 **36** 1637-44

[13] Basu S, 2012 Tensile Deformation of Fibers Used in Textile Industry *Agilent Technologies Application Note*

[14] Mishra S, Mohanty A K, Drzal L T, Misra M, Parija S, Nayak S K, Tripathy S S 2003 *Compos.* Sci. Technol. 63 1377-1385

[15] Matthews F L, Rawlings R D 1994 Composite Materials: Engineering and Science, Chapman and Hall (London)

[16] Liu F P, Wolcott M P, Gardner D J, Rials T G 1995 Composite Interfaces 2 419-432

IOP Conf. Series: Materials Science and Engineering **48** (2013) 012009 doi:10.1088/1757-899X/48/1/012009

Degradation of elastic properties of non-uniformly damaged composite laminates

J Varna[1], M S Loukil[1,2], Z Ayadi[2]
[1] Division of Materials Science, Luleå University of Technology, SE-971 87 Luleå, Sweden
[2] Institut Jean Lamour, EEIGM 6 Rue Bastien Lepage, F-54010 Nancy Cedex, France

E-mail: mohamed.loukil@ltu.se

Abstract. Stiffness reduction model for laminates with non-uniformly distributed intralaminar cracks is presented and used to analyze the effect of non-uniformity and the accuracy of predictions based on uniform spacing. The values of the normalized crack opening displacement (COD) as affected by the presence of other cracks are used to calculate the axial modulus of cross-ply laminates with cracks. The COD is calculated using finite element method (FEM), considering two closest neighbors of the crack and using the smallest versus the average crack spacing ratio as non-uniformity parameter. Assuming uniform spacing the axial modulus reduction is overestimated. A "double-periodic" approach is presented to calculate the COD of a crack in a non-uniform case as the average of two solutions for periodic crack systems.

1. Introduction

Intralaminar cracking in cross-ply laminates is the first mode of damage. The crack plane is orthogonal to the laminate midplane, they run parallel to fibers in the layer and often cover the whole thickness and width of the layer.

Usually the extent of cracking (number of cracks and distance between them) is characterized by average crack spacing and crack density (cracks/mm). Most of the existing stiffness models, for example, [1-7] use this assumption. However, the crack distribution in the layer may be highly non-uniform. The reason is that the transverse failure strength along the longitudinal direction of the layer has statistical distribution. The discussion in this paper is focused on the possible inaccuracy introduced in laminate stiffness prediction by using assumption of uniform spacing. We will develop a theoretical framework and analyze two cases: a) the system of cracks is "non-interactive" in average (low crack density) but some cracks are close to each other and interact; b) the crack density is high and all cracks interact even "in average".

There are only a few studies where the effect of non-uniformity is addressed, see for example [5]. In [5] hypothesis was introduced that for a non-uniformly cracked laminate, the deformation field in the "element" between two neighboring ply cracks separated by a distance l_k is identical to that in a uniformly cracked laminate where the crack spacing is l_k. Then, for example, the axial strain of the whole Representative Volume Element (RVE) can be calculated by the "rule of mixtures" of average strains of "elements". Good agreement of this approach with another high accuracy semi-analytical methodology applied to the RVE was demonstrated.

The average value of the axial stress change between two cracks is proportional to the average crack opening displacement COD and crack sliding displacements (CSD) normalized with far field stress [6]. Therefore the damaged laminate stiffness reduction depends on density of cracks and these parameters: average COD and CSD as expressed in the GLOB-LOC model [1, 7]. These two rather robust parameters depend on the normalized distance to neighboring cracks. Therefore for non-uniform crack distribution they are different for each individual crack. The values of COD and CSD in

the commonly assumed uniform crack distribution case correspond to average spacing between cracks and are different than the calculated average over COD's and CSD's of all individual cracks.
We will generalize the previously developed GLOB-LOC model [1] to non-uniform spacing case. Parametric analysis of the effect of geometrical non-uniformity in terms of COD and the laminate axial modulus will be performed for $[0_n/90_m]_s$ cross-ply laminates with cracks in 90-layers. To simplify stiffness calculations for an arbitrary non-uniform distribution, a routine allowing determination of COD for any crack as a sum of solutions for two periodic systems of cracks, will be formulated: one solution is for periodic system with spacing as on the "+" side of the crack and another one for a periodic system with spacing as on the "-"side of the crack.

2. Elastic constants of damaged symmetric cross-ply laminates with intralaminar cracks

2.1. Distance between cracks

We consider RVE of a layer (index omitted) with M cracks as schematically shown in Fig 1. Crack with index m has two neighbors located at different distances l_{m-1} and l_m from this crack. The RVE length is L, the average distance between cracks (spacing) is l_{av}, the crack density is ρ

$$L=\sum_{m=0}^{M-1} l_m \qquad l_{av}=\frac{L}{M} \qquad \rho=\frac{1}{l_{av}} \tag{1}$$

We will use the distance between cracks normalized with respect to the layer thickness t

$$l_{mn}=\frac{l_m}{t} \qquad m=0,1,\ldots..M-1 \qquad l_{avn}=\frac{l_{av}}{t} \tag{2}$$

We denote u^m_{2an}, u^m_{1an} the average normalized COD and CSD of the m-th crack. Obviously

$$u^m_{2an}=u^m_{2an}\left(l_{(m-1)n}, l_{mn}\right) \qquad u^m_{1an}=u^m_{1an}\left(l_{(m-1)n}, l_{mn}\right) \tag{3}$$

If $l_m > l_{m-1}$ the displacements on the "-" (right) side will be larger than on the "+" (left) side.

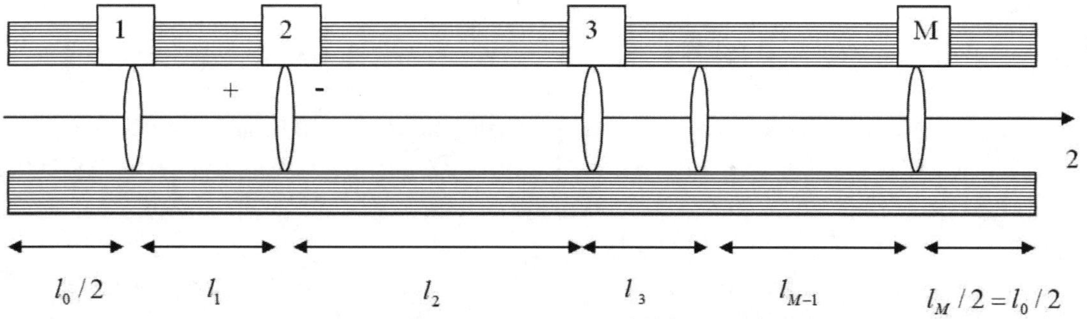

Figure 1. Non-uniform distribution of M cracks in damaged layer shown in its local coordinate system

2.2. Laminate's stress-strain relationship with Valulenko-Kachanov tensor

We consider symmetric N- layer laminate. The k-th layer of the laminate has thickness t_k, fiber orientation angle with respect to the global x-axis θ_k and stiffness matrix $[Q]$ in the material symmetry axes, calculated from elastic constants $E_1, E_2, G_{12}, \nu_{12}$. The total thickness of the laminate, $h=\sum_{k=1}^{N} t_k$. Each layer can be damaged and the crack density in a layer, ρ_k is calculated using (1) where the average distance between cracks l^k_{av} is measured transverse to the fiber direction in the k-th layer. Dimensionless crack density ρ_{kn} in the layer is introduced as

$$\rho_{kn} = t_k \rho_k \tag{4}$$

Using divergence theorem, it is easy to show [8] that for stress states that satisfy equilibrium equations the average stress applied to external boundary is equal to volume averaged stress. For laminated composites with applied average in-plane stress vector, $\{\sigma_0\}^{LAM}$ this equality can be written as

$$\{\sigma_0\}^{LAM} = \{\bar{\sigma}\}^a = \sum_{k=1}^{N} \{\bar{\sigma}\}_k^a \frac{t_k}{h} \tag{5}$$

In (5) the volume average is calculated expressing the integral over the laminate volume as a sum of volume integrals over N layers. Upper index a is used to indicate volume averages. The bar over stresses indicates global coordinates. In global coordinates the Hook's law in averaged form is the same as in any point

$$\{\bar{\sigma}\}_k^a = [\bar{Q}]_k \{\varepsilon\}_k^a \tag{6}$$

Substituting (6) in (5) and using the relationship (see [1])

$$\begin{Bmatrix} \varepsilon_x^a \\ \varepsilon_y^a \\ \gamma_{xy}^a \end{Bmatrix}_k = \begin{Bmatrix} \varepsilon_x \\ \varepsilon_y \\ \gamma_{xy} \end{Bmatrix}^{LAM} + \begin{Bmatrix} \beta_x \\ \beta_y \\ 2\beta_{xy} \end{Bmatrix}_k \tag{7}$$

we obtain

$$\{\sigma_0\}^{LAM} = [Q]_0^{LAM} \{\varepsilon\}^{LAM} + \sum_{k=1}^{N} [\bar{Q}]_k \frac{t_k}{h} \{\bar{\beta}\}_k \tag{8}$$

In (8) $\{\bar{\beta}\}$ is the Vakulenko-Kachanov tensor [8] written in Voigt notation. In Cartesian coordinates

$$\beta_{ij} = \frac{1}{V} \int_{S_C} \frac{1}{2}\left(u_i n_j + u_j n_i\right) dS \tag{9}$$

Integration in (9) involves the total surface S_c of all cracks in the layer, u_i are displacements of the points on the crack surface, n_i is outer normal to the crack surface, V is the volume of the layer. $\{\bar{\beta}\}$ in (8) represents the effect on stiffness of the crack face displacements (COD and CSD). Since β_{ij} and strain are tensors for both of them we have the same transformation expressions between local and global coordinates

$$\{\bar{\beta}\}_k = [T]_k^T \{\beta\}_k \tag{10}$$

2.3. Incorporation of COD and CSD in Valulenko-Kachanoy tensor (local coordinates)

The cracked layer is considered in its local coordinates with indexes 1, 2 and 3 corresponding to longitudinal, transverse and thickness directions. Index denoting k-th layer is omitted to simplify explanation in this section. In local coordinates the normal vector to the two faces of crack surface has coordinates

$$n_1 = n_3 = 0 \qquad\qquad n_2 = \pm 1 \tag{11}$$

Using the definition (9) for β_{ij} we see that it contains only two non-zero elements: β_{12} and β_{22}

$$\beta_{22} = \frac{1}{Lt} \sum_{m=1}^{M} \int_{-t/2}^{+t/2} \left[u_2^{m+}(z) - u_2^{m-}(z)\right] dz \qquad\qquad \beta_{12} = \frac{1}{Lt} \sum_{m=1}^{M} \int_{-t/2}^{+t/2} \frac{1}{2}\left[u_1^{m+}(z) - u_1^{m-}(z)\right] dz \tag{12}$$

In (12) t is the cracked layer thickness, $u_1^m(z)$ and $u_2^m(z)$ are sliding and opening displacements of the m-th crack, symbol + or – denotes the particular crack face as defined in 2.1. If the crack density is high the crack face displacements depend on the distance between cracks. We define the average value of the opening and sliding on each crack surface

$$u_{1a}^{m+} = \frac{1}{t}\int_{-t/2}^{+t/2}\left[u_1\left(t/2\right) - u_1^{m+}(z)\right]dz \qquad u_{1a}^{m-} = \frac{1}{t}\int_{-t/2}^{+t/2}\left[u_1^{m-}(z) - u_1\left(t/2\right)\right]dz \tag{13}$$

$$u_{2a}^{m+} = \frac{1}{t}\int_{-t/2}^{+t/2}\left[u_2\left(t/2\right) - u_2^{m+}(z)\right]dz \qquad u_{2a}^{m-} = \frac{1}{t}\int_{-t/2}^{+t/2}\left[u_2^{m-}(z) - u_2\left(t/2\right)\right]dz \tag{14}$$

The average value of the average opening and the average sliding of both faces is

$$u_{1a}^{m} = \frac{1}{2}\left(u_{1a}^{m+} + u_{1a}^{m-}\right) \qquad u_{2a}^{m} = \frac{1}{2}\left(u_{2a}^{m+} + u_{2a}^{m-}\right) \tag{15}$$

Using (13) , (14) and (15) the expressions for β_{12} and β_{22} are

$$\beta_{12} = -\frac{1}{L}\sum_{m=1}^{M} u_{1a}^{m}\left(l_{(m-1)n}, l_{mn}\right) \qquad \beta_{22} = -\frac{2}{L}\sum_{m=1}^{M} u_{2a}^{m}\left(l_{(m-1)n}, l_{mn}\right) \tag{16}$$

We indicate in (16) that the displacements will be mostly affected by normalized distances to the two closest neighboring cracks. Expressions (16) can be rewritten in terms of crack density and average (over all cracks) displacements

$$\beta_{12} = -\rho u_{1a} \qquad \beta_{22} = -2\rho u_{2a} \tag{17}$$

$$u_{1a} = \frac{1}{M}\sum_{m=1}^{M} u_{1a}^{m}\left(l_{(m-1)n}, l_{mn}\right) \qquad u_{2a} = \frac{1}{M}\sum_{m=1}^{M} u_{2a}^{m}\left(l_{(m-1)n}, l_{mn}\right) \tag{18}$$

We normalize u_{1a} and u_{2a} with respect to CLT stresses σ_{20}, σ_{120} and ply thickness t

$$u_{1an} = \frac{G_{12}}{t\sigma_{120}}\frac{1}{M}\sum_{m=1}^{M} u_{1a}^{m}\left(l_{(m-1)n}, l_{mn}\right) \qquad u_{2an} = \frac{E_2}{t\sigma_{20}}\frac{1}{M}\sum_{m=1}^{M} u_{2a}^{m}\left(l_{(m-1)n}, l_{mn}\right) \tag{19}$$

For normalized values the expressions for β_{ij} in (17) are slightly modified. They can be presented in the following matrix form

$$\{\beta\} = \begin{Bmatrix} 0 \\ \beta_{22} \\ 2\beta_{12} \end{Bmatrix} = -\frac{\rho_n}{E_2}[U]\begin{Bmatrix} \sigma_{10} \\ \sigma_{20} \\ \sigma_{120} \end{Bmatrix} \qquad [U] = 2\begin{bmatrix} 0 & 0 & 0 \\ 0 & u_{2an} & 0 \\ 0 & 0 & \frac{E_2}{G_{12}}u_{1an} \end{bmatrix} \tag{20}$$

In (20) ρ_n is normalized crack density in the layer defined by (4).

2.4. Homogenized stiffness of the damaged laminate

Using the classical laminate theory (CLT), the stress $\{\sigma_0\}_k$ in the k-th layer in local coordinates can be expressed through the applied laminate stress $\{\sigma_0\}^{LAM}$ as follows

$$\begin{Bmatrix} \sigma_{10} \\ \sigma_{20} \\ \sigma_{120} \end{Bmatrix}_k = [T]_k\begin{Bmatrix} \sigma_{x0} \\ \sigma_{y0} \\ \sigma_{xy0} \end{Bmatrix}_k = [T]_k\left[\bar{Q}\right]_k\begin{Bmatrix} \varepsilon_{x0} \\ \varepsilon_{y0} \\ \gamma_{xy0} \end{Bmatrix}^{LAM} = [T]_k\left[\bar{Q}\right]_k[S]_0^{LAM}\begin{Bmatrix} \sigma_{x0} \\ \sigma_{y0} \\ \sigma_{xy0} \end{Bmatrix}^{LAM} \tag{21}$$

Substituting (20) with (21) in (10) and further in (8) we obtain after arranging the result in the following form

$$\{\sigma_0\}^{LAM} = [Q]^{LAM}\{\varepsilon\}^{LAM} \tag{22}$$

with the following expressions for the stiffness and the compliance matrices of the damaged laminate

$$[Q]^{LAM} = \left([I] + \sum_{k=1}^{N} \rho_{kn}\frac{t_k}{h}[P]_k[S]_0^{LAM}\right)^{-1}[Q]_0^{LAM} \tag{23}$$

$$[S]^{LAM} = [S]_0^{LAM}\left([I] + \sum_{k=1}^{N} \rho_{kn}\frac{t_k}{h}[P]_k[S]_0^{LAM}\right) \tag{24}$$

The compliance matrix of the undamaged laminate is $[S]_0^{LAM} = \left([Q]_0^{LAM}\right)^{-1}$. In (23) and (24) the matrix-function $[P]_k$ for a layer with index k is defined as

$$[P]_k = \frac{1}{E_2}[\overline{Q}]_k[T]_k^T[U]_k[T]_k[\overline{Q}]_k \tag{25}$$

The involved matrices $[T]_k$ and $[\overline{Q}]_k$ are defined according to CLT, upper index T denotes transposed matrix and bar over stiffness matrix indicates that it is written in global coordinates. The influence of cracks in k-th layer is represented by matrix $[U]_k$ given by (20). Index k is added because the matrix depends on the constraint of surrounding layers and therefore u_{2an}, u_{1an} may be different for different layers. Elements of this matrix u_{2an}^k, u_{1an}^k are calculated, see (19), using normalized and averaged crack face opening (COD) and sliding displacements (CSD) of all cracks as affected by varying spacing between them.

2.5. Elastic modulus of balanced laminates with cracks in 90-layer

In case of balanced laminates with damage in 90-layers only, expressions for $[P]_k$ and for $[S]^{LAM}$ have been obtained calculating analytically the matrix products in (23) - (25). For example, the obtained expression for laminate normalized axial modulus is

$$\frac{E_x}{E_x^0} = \frac{1}{1 + 2\rho_{90n}\frac{t_{90}}{h}u_{2an}^{90}c} \qquad c = \frac{E_2}{E_x^0}\left(\frac{1-\nu_{12}\nu_{xy}^0}{1-\nu_{12}\nu_{21}}\right)^2 \tag{26}$$

Subscript 90 is used to indicate 90-layer. The quantities with lower index x,y are laminate constants, quantities with additional upper index 0 are undamaged laminate constants and (19) has to be used to calculate u_{2an}^{90}. In the case of uniform crack distribution all CODs in (19) are equal. In this paper we analyze damaged laminates with axial modulus given by (26) and therefore only COD is of interest.

3. Results and discussions

3.1. Formulation of calculation examples

The effect of the non-uniform crack distribution on u_{2an}^{90} was analyzed using FEM for damaged $[0_n / 90_8]_S$ laminates (n=1,8) at fixed dimensionless crack density ρ_{90n}, see Fig 2 where the repeating "super-element" with two cracks is shown. To characterize the non-uniformity of the spatial distribution parameter K is introduced

$$K = \frac{l_0}{l_{av}} \tag{27}$$

The same value of K holds for normalized spacing. For uniform crack distribution $K = 1$. The average normalized crack spacing, l_{av} was kept constant $l_{avn} = \frac{1}{2}(l_{0n} + l_{1n})$.

CF/EP and GF/EP composites with constants given in Table 1 were analyzed. All plies are considered to be transversally isotropic with $E_2 = E_3$, $G_{12} = G_{13}$ and $\nu_{12} = \nu_{13}$ (1 is the longitudinal direction and 2 and 3 are the transverse directions of the fiber). All results are presented in terms of normalized crack spacing and normalized crack density. Stiffness depends on layer thickness ratio, not on the absolute value of ply thickness.

The commercial code ABAQUS was used to find the solution for the "super-element" in Fig. 2. 3D continuum elements (C3D8) 8-node linear brick were used. The same mesh with total number of 86400 elements was used in each FE model. To the right boundary $x = 0$ of the model a given constant displacement in x-direction corresponding to 1% average strain was applied keeping a symmetry boundary condition at the left side. The top surface was free of tractions. On the front edge (y=0) and the edge y=w (w is the width of the laminate) coupling conditions were applied (U_y =unknown constant). To cover large variation in elastic constants both CF/EP and GF/EP composites with constants given in Table 1 were analyzed. Studying the effect of non-uniform distribution the normalized spacing l_{0n}, see Fig. 2, was used as a parameter which was lower or equal to the average spacing.

Table 1. Composite elastic constants used in simulations.

Material	E_1(GPa)	E_2(GPa)	ν_{12}	ν_{23}	G_{12} (GPa)	G_{23} (GPa)
GF/EP	45	15	0.3	0.4	5	5.36
CF/EP	150	10	0.3	0.4	5	3.57

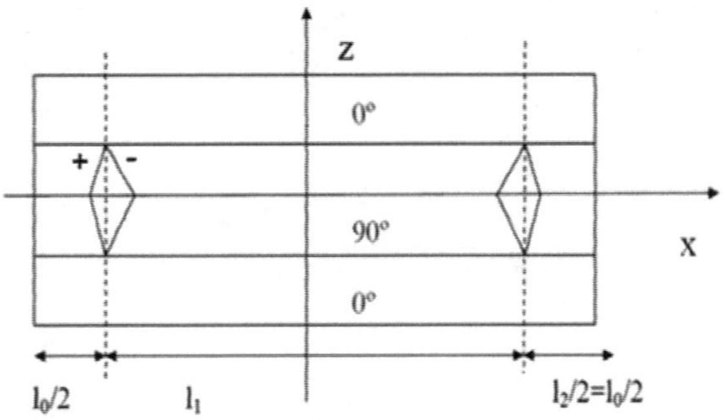

Figure 2. "Super-element" model for COD studies with non-uniformly cracked 90-layers.

3.2. COD parametric analysis at low crack density

Results are presented for average normalized spacing $l_{avn} = 10$ ($\rho_{90n} = 0.1$). The profiles of normalized crack face displacements $u_{2n}^{-}(z_n)$, $u_{2n}^{+}(z_n)$ along the thickness coordinate $z_n = z\frac{2}{t_{90}} + 1$ are shown in Fig. 3. The "+" face of the crack has smaller displacements than the "-" face and the difference is larger when the l_{0n} is smaller than 1 (the neighboring crack to the left is very close). The neighbor to the "-" face is at larger distance than the average spacing and therefore the displacement profile is almost unaffected. For the same geometry the CODs in CF composites are always significantly smaller than the CODs in GF. The CODs are significantly smaller when the relative thickness of the neighboring layer is higher. This effect is more pronounced for GF composite.

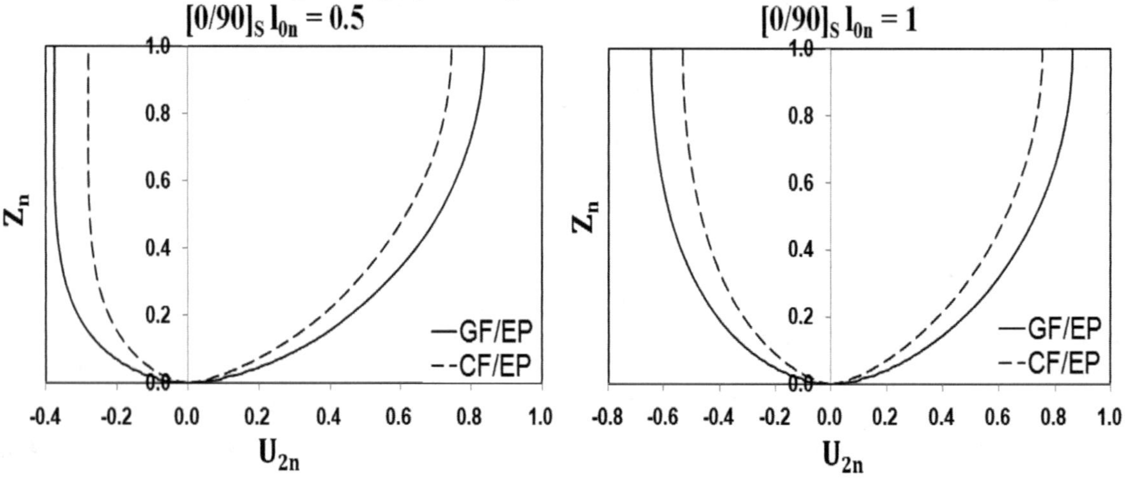

Figure 3. COD profiles of cracks in [0/90]s laminate with normalized crack density $\rho_{90n} = 0.1$

Using crack face displacements the average normalized CODs, u_{2an}^{90} are calculated by numerical integration. The obtained dependence on the non-uniformity parameter K is shown in Fig. 4. The u_{2an}^{90} is larger if the spacing is uniform. However, the effect is negligible for $K > 0.2$ ($l_0 > 2t_{90}$).

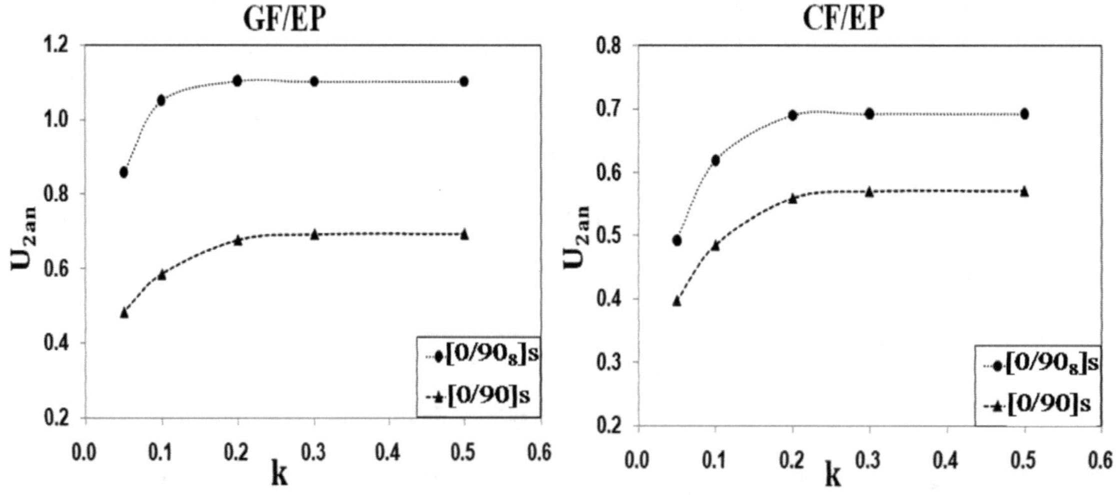

Figure 4. Effect of non-uniform spacing on COD of cracks in cross-ply laminates $\rho_{90n} = 0.1$

IOP Conf. Series: Materials Science and Engineering **48** (2013) 012009 doi:10.1088/1757-899X/48/1/012009

3.3. Approximate COD determination from periodic solutions

The average normalized COD, u_{2an}^{90} of a crack in a layer with non-uniform crack distribution can be found considering separately the average normalized COD of the"-" face of the crack and"+" of the crack.

$$u_{2an}^{90} = \frac{1}{2}\left(u_{2an}^{+} + u_{2an}^{-}\right) \tag{28}$$

In this section the following hypothesis will be inspected:
"The opening of a given crack face ("+" or "-" in Fig. 2) depends only on the distance (l_{0n} or l_{1n}) to the closest neighboring crack on that side. It can be determined considering the corresponding region between these two cracks as a periodic element."
According to this "double-periodic" approach u_{2an} is obtained as

$$u_{2an}^{90} \approx u_{2an}^{p}, \qquad u_{2an}^{p} = \frac{1}{2}\left(u_{2an}^{p-} + u_{2an}^{p+}\right) \tag{29}$$

The two values u_{2an}^{p+}, u_{2an}^{p-} are solutions of the two periodic models. This hypothesis is equivalent to statement that in Fig. 2 symmetry conditions on the plane $x = \pm l_1/2$ can be applied. If the "double-periodic" approach is accurate enough, the u_{2an} for any crack location with respect to other cracks could be calculated from a master curve for uniform crack distribution. This curve, which is expression of u_{2an} as a function of crack spacing in a layer with uniformly distributed cracks, would be used twice to read the u_{2an}^{p+}, u_{2an}^{p-} values of the left and the right face of the crack.
In order to check the accuracy and validity of the "double-periodic" assumption, the u_{2an}^{90} for each value of non-uniformity parameter was calculated in two different ways: a) directly applying FEM to the non-uniform geometry; b) applying FEM two times and using (29).
From Fig. 5 where displacement profiles according to a) and b) are presented we conclude that the trends in the double-periodic approach are described correctly but the values of face displacements are not accurate. On the left face where the interaction is strongest the u_{2n}^{p+} is too small but on the right face, where the next crack is further away, u_{2n}^{p-} is too large.

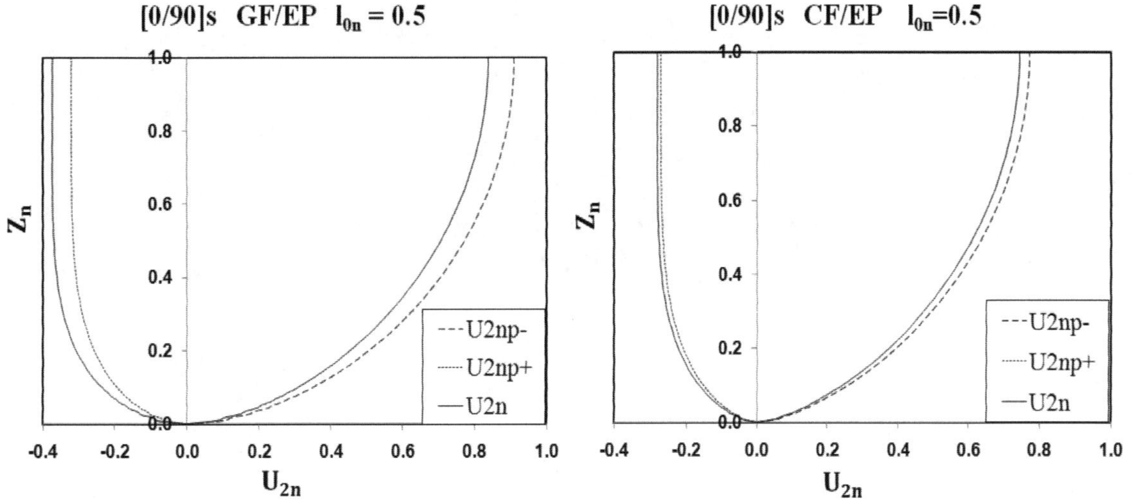

Figure 5. Calculated COD profiles of cracks in [0/90]s laminate with normalized crack density $\rho_{90n} = 0.1$

IOP Conf. Series: Materials Science and Engineering **48** (2013) 012009 doi:10.1088/1757-899X/48/1/012009

However, the u_{2an}^{p} given by (29) is requested for stiffness predictions and not the value for each face separately. The values of u_{2an} and u_{2an}^{p} can be compared using results presented in Table 2 for all lay-ups, materials and non-uniformity parameter values. A very good agreement between values exists for all cases which validates the use of the "double-periodic" hypothesis.

Table 2. Average normalized COD of cracks from FEM and from "double-periodic" approach $\rho_{90n} = 0.1$

	[0/90$_8$]s				**[0/90]s**			
	GF/EP		**CF/EP**		**GF/EP**		**CF/EP**	
K	u_{2an}	u_{2an}^{p}	u_{2an}	u_{2an}^{p}	u_{2an}	u_{2an}^{p}	u_{2an}	u_{2an}^{p}
0.50	1.1027	1.1030	0.6927	0.6927	0.6941	0.6943	0.5721	0.5722
0.30	1.1027	1.1030	0.6928	0.6931	0.6928	0.6935	0.5712	0.5716
0.20	1.1039	1.1045	0.6905	0.6907	0.6771	0.6786	0.5600	0.5607
0.10	1.0510	1.0537	0.6189	0.6230	0.5860	0.5910	0.4857	0.4894
0.05	0.8588	0.8811	0.4927	0.5091	0.4841	0.4890	0.3977	0.4035

4. Elastic Modulus prediction and validation with FEM

The effect of the non-uniform crack distribution on axial modulus of cross-ply laminates is shown in Fig 6. All results are for the same normalized crack density $\rho_{90n} = 0.1$. The normalized axial modulus of the laminate is calculated in three different ways:

a) Calculating the average applied stress using FEM and then using definition of $E_x = \dfrac{\sigma_x}{\varepsilon_x}$;

b) Applying (26) and using for u_{2an}^{90} values of u_{2an} obtained from FEM and presented in Table 2;

c) Applying (26) and using for u_{2an}^{90} values of u_{2an}^{p} obtained from "double-periodic" approach presented in Table 2;

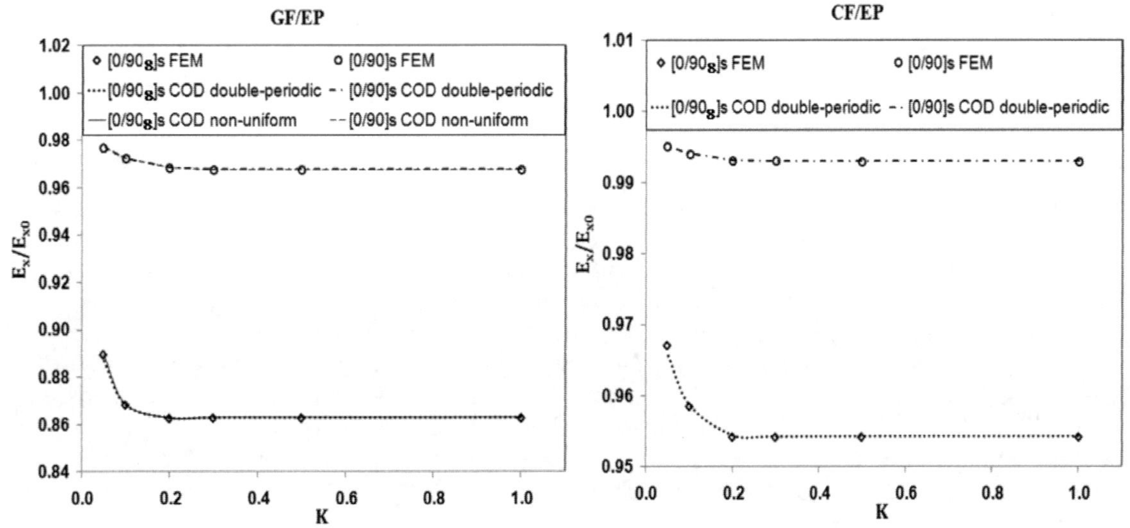

Figure 6. Effect of non-uniform crack distribution on axial modulus of cross-ply laminates $\rho_{90n} = 0.1$

The elastic modulus of the RVE with two non-equidistant cracks calculated directly from FEM coincides with the elastic modulus for this RVE calculated using (26) with u_{2an} input from the same FEM solution (called " COD non-uniform" in Fig.6). Since (26) is an exact analytical expression this result was expected.

On other hand, the "double-periodic" approach values practically coincide with the FEM values, proving the accuracy and potential of this approach for simulation of systems with multiple non-uniformly spaced cracks.

For the used crack density and all investigated materials and lay-ups the axial modulus reduction is the highest if cracks have uniform distribution. The effect of non-uniform distribution of internal cracks on laminate modulus can be neglected if the non-uniformity parameter $K > 0.2$.

Similar calculations as described above were performed for higher crack density $\rho_{90n} = 0.5$ Results are presented in Fig. 7.

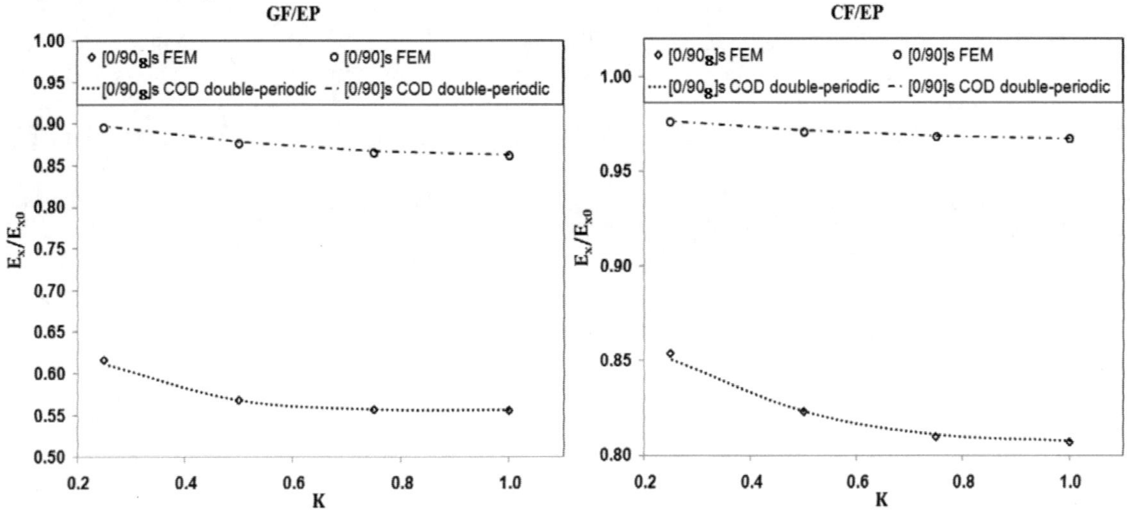

Figure 7. Effect of non-uniform crack distribution on axial modulus of cross-ply laminates, $\rho_{90n} = 0.5$

The effect of non-uniform distribution is even larger than at low crack density. In contrast to low crack density case, there is no plateau region in Fig. 7. The "double-periodic" approach at high crack density is still highly accurate.

5. Conclusions

Elastic properties of damaged symmetric cross-ply laminates can be predicted and the effect of non-uniform crack distribution can be analyzed using the derived exact expressions. However, the COD and the CSD in these expressions have to be calculated solving numerically the stress distribution problem for the RVE or using approximate solutions. The COD, CSD and the number of cracks per unit length in the layer are governing the laminate properties reduction.

This model was applied to cross-ply laminates with cracks in 90-layers. The effect of the non-uniform crack distribution on the damaged cross-ply laminate axial modulus was analyzed numerically. COD values needed as an input in the model were calculated using FEM and stiffness calculations were performed for GF/EP as well as CF/EP laminates with low and also with high crack density.

The same trend was found for all crack densities and lay-ups: assuming uniform crack distribution the damaged laminate modulus is underestimated.

An approximate "double-periodic" approach was proposed stating that the COD of a crack with different distances to the closest neighbors can be calculated as the average of two solutions for equidistant cracks. It was shown numerically for cross-ply laminates that very accurate COD values for cracks with non-uniform spacing and elastic modulus values can be obtained using this approach.

6. References

[1] Lundmark P and Varna J 2005 Constitutive relationships for laminates with ply cracks in in plane loading *Int. J. of Damage Mech.* **14** **(3)** 235-61

[2] Hashin, Z 1985 Analysis of Cracked Laminates: A Variational Approach *Mech of Mat.* **4** **(2)** 121-136

[3] Talreja R 1994 Damage characterization by internal variables, In: Pipes, R.B., Talreja, R. (eds), Damage Mechanics of Composite Materials, Composite Materials series **9** 53-78 Elsevier, Amsterdam

[4] Kashtalyan M and Soutis C 2005 Analysis of Composite Laminates with Intra- and Interlaminar Damage, Progress in Aerospace Sciences **41** **(2)** 152-173

[5] McCartney LN and Schoeppner GA 2000 Predicting the Effect of Non-uniform Ply Cracking on the Thermo-elastic Properties of Cross-ply Laminates *Comp. Sc. and Tech.* **62** 1841-56

[6] Varna J, Krasnikovs A, Kumar R and Talreja R 2004 A Synergistic Damage Mechanics Approach to Viscoelastic Response of Cracked Cross-ply Laminates *Int. J. of Damage Mech.* **13** 301-334

[7] Lundmark P and Varna J 2006 Crack face sliding effect on stiffness of laminates with ply cracks *Comp. Sc. Tech.* **66** 1444–54

[8] Vakulenko AA and Kachanov ML 1971 Continuum Model of Medium with Cracks, Mekhanika Tverdogo Tela (Mechanics of Solids) **4** 159-166

Influence of hydrogen environment on fatigue crack growth in forged Ti-6Al-4V: fractographic analysis

R Gaddam[1], R Pederson[1], M Hörnqvist[2], M-L Antti[1]
[1]Division of Materials Science, Luleå University of Technology, 97187 Luleå, Sweden

[2]Department of Applied Physics, Chalmers University of Technology, 41296 Göteborg, Sweden

E-mail: raghuveer.gaddam@ltu.se

Abstract. The influence of high-pressure gaseous hydrogen environment (15 MPa) on the fatigue crack growth in forged Ti-6A1-4V at room temperature is investigated. It is observed that the fatigue crack growth (FCG) rate is fluctuating at $20 \leq \Delta K \leq 26$ MPa√m, and increase drastically at $\Delta K > 26$ MPa√m in hydrogen environment. The effect of hydrogen on the FCG rate is dependent on the stress intensity level (ΔK). Detailed fractographic analysis of the fracture surfaces is performed at different ΔK using field emission scanning electron microscope (FE-SEM). The differences in appearance of fracture surfaces in air and hydrogen are discussed.

1. Introduction

Titanium and its alloys are widely used in aerospace industry owing to their combination of high specific strength (i.e. strength/density), good fatigue strength and excellent corrosion resistance [1-2]. Among the many, Ti-6Al-4V (referred hereafter as Ti-64) is the principal alloy that is used in aeroengines up to 300°C [2]. Despite its exceptional properties, it has very limited usage in space applications. One of the major concerns is its incompatibility with hydrogen [3]; which is primarily used as fuel or coolant in rocket engines. Previous studies [4-9] have shown that presence of hydrogen in an environment (i.e. gaseous hydrogen) degraded mechanical properties such as ductility, fracture toughness and fatigue strength at ambient temperature.

Ti-64 is an aerospace alloy; the components made of this alloy are subjected to cyclic loads in service where the fatigue properties are of particular importance. If the alloy is intended for use in space applications, increased understanding of the effect of hydrogen on fatigue properties is required. The aim of the present work was to perform a detailed fractographic analysis to understand the effect of hydrogen environment (15 MPa) on fatigue crack growth resistance of forged Ti-64 at room temperature.

2. Materials and methods

2.1. Materials and environment

The material investigated is a Ti-64 alloy in forged condition, which has a bi-modal microstructure i.e. consisting of primary α and transformed β grains (see figure 1(a)). Transformed β is represented by α/β structure containing α colonies and α laths (figure 1(b)). Table 1 shows the quantitative information of these microstructural features. The environments used were ambient air and high-pressure gaseous hydrogen ($\approx$ 99.995% purity), where the desired pressure (15 MPa) was obtained by pressurizing the test chamber prior to testing.

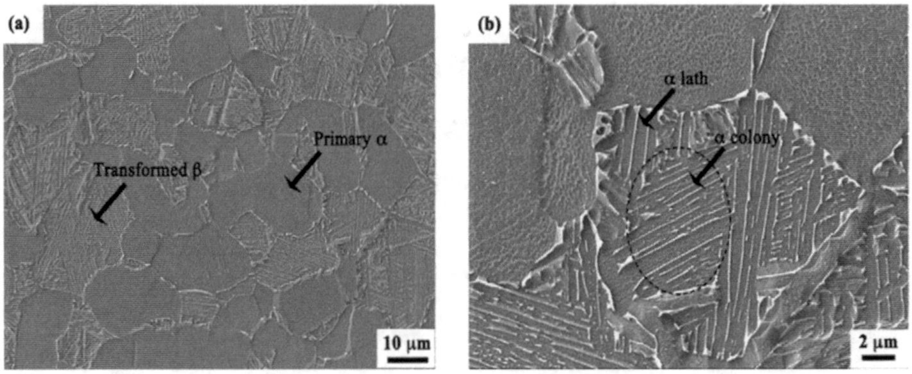

Figure 1. (a) Scanning electron micrographs of forged Ti-64 revealing the bi-modal microstructure with primary α and transformed β grains, (b) Magnified image of transformed β in (a) showing the α/β structure. Grey area represents α phase and white shows β phase.

Table 1. Measurements of microstructural features in forged Ti-64 alloy.

Volume fraction of primary α (%)	Volume fraction of transformed β (%)	Size of primary α (μm)	Size of transformed β (μm)	Size of α colony (μm)	Alpha lath thickness (μm)
42.5	57.5	15 ± 3.87	18 ± 5.68	10 ± 2.68	0.52 ± 0.15

2.2. Fatigue crack growth testing

Fatigue crack growth (FCG) testing was conducted using a servohydraulic test rig on one sample each in ambient air and hydrogen environment at room temperature. The test specimens used were Kb-type, which have rectangular cross sections of 4.3 x 10.2 mm in the gauge section (dotted box in figure 2(a)). An initial starter notch of nominal depth of 0.1 mm and total width of 0.2 mm was introduced using electric discharge machining. The specimens along with the grips were enclosed in the autoclave as shown in figure 2(b), and the loads measured were corrected for internal autoclave pressure and piston seal friction. The specimens were pre-cracked under fatigue loading in air at room temperature, at a frequency of 10 Hz with a stress ratio $R = \sigma_{min}/\ \sigma_{max} = 0$, to obtain an initial crack length of ~ 0.5 mm. After pre-cracking, FCG testing was carried out by loading the specimens uniaxially with $R = 0$ and a frequency of 0.5 Hz, with triangular waveform by using the applicable parts of ASTM E647-08 [10] and ASTM E740-03 [11]. In these tests, the total crack lengths were measured by the direct current potential drop (PD) technique according to ASTM E 647 using an experimentally obtained calibration function to relate the PD signal to crack length. The tests were interrupted at an approximate total crack length of 2.6 mm. After the testing, the specimen tested in hydrogen environment was heat treated at about 150°C to mark the final crack front on the fracture surface.

The stress intensity factor range, ΔK, was calculated according to ASTM E740 as:

$$\Delta K = f^* \Delta\sigma^* \sqrt{(\pi a)} \tag{1}$$

where $\Delta\sigma$ is the applied stress range, a is the crack length and ı is a geometric factor based on specimen and crack geometry. Here, ΔK was calculated for the deepest point of the crack assuming a semi-circular shape, which was verified from the fracture surfaces. As the applied stress range was different for the tests in air and hydrogen (450 and 530 MPa, respectively), equation (1) was also used to correlate the crack lengths (ı) between the tests, in order to compare fractographic results obtained at the same value of ΔK.

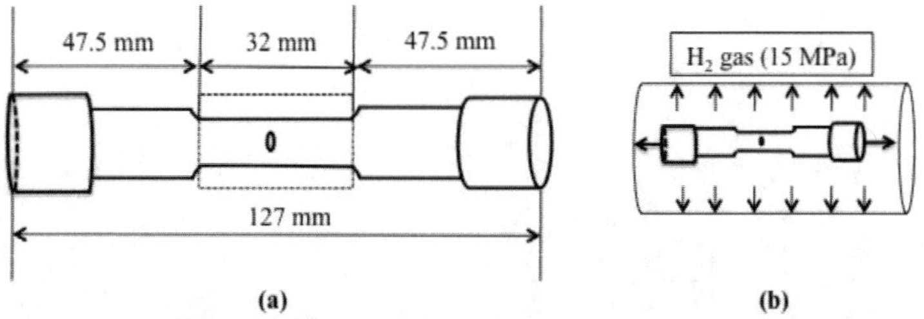

Figure 2. (a) Schematic diagram showing the dimensions of the Kb type specimen used in testing and (b) schematic illustration of testing in 15 MPa H_2 gas in closed autoclave.

2.3. Fractography

Fractographic analysis of fracture surfaces after the FCG tests were conducted using a field emission gun scanning electron microscope (FE-SEM) from Carl Zeiss (MERLIN® FE-SEM). The analysis were performed using secondary electrons with an accelerating voltage of 3 kV and probe current of 1 nA. During fractographic observations, attention was paid on specific regions showing distinct changes in the fracture surface features along the crack propagation direction.

3. Results and discussion

3.1. Fatigue crack growth (FCG) rate

FCG rate curve of forged Ti-64 in ambient air and gaseous hydrogen environment is shown in figure 3(a). It is expressed by showing a relation between fatigue crack growth rate (da/dN in mm/cycle) and stress intensity factor range (ΔK in MPa√m). Figure 3(b) shows the ratio of FCG rate change as a function of stress intensity. It was determined at various stress-intensity factor ranges (ΔK) by dividing the crack growth rate in hydrogen $(da/dN)_{hydrogen}$ by that obtained in air $(da/dN)_{air}$.

The results from FCG testing in air and hydrogen environment can be described by dividing the FCG rate curve into three distinct regions (see figure 3(b)). These regions are also reflected in the changes seen on the macroscopic fracture surface (see figure 4(b)). In the first region, denoted FCG-1 ($\Delta K \leq 20$ MPa√m, $a \leq 1.2$ mm), the da/dN ratio values are found to be nearly 1. It means that the FCG rate in hydrogen environment remains unaffected at low stress intensity range, and showing a stable crack growth similar to the one found in air. In the second region, denoted FCG-2 ($20 \leq \Delta K \leq 26$ MPa√m, $1.2 \leq a \leq 1.5$ mm), the da/dN ratio values are lower, on average, which shows that the FCG rate in hydrogen environment is lower than in air at intermediate stress intensity range. It is also observed that there are fluctuations in the da/dN values up to $\Delta K \approx 26$ MPa√m. These fluctuations show that crack growth is not stable or crack is temporarily arrested in hydrogen environment. Finally, in the third region, denoted as FCG-3 ($\Delta K > 26$ MPa√m, $a \geq 1.5$ mm), the da/dN ratio values drastically increase to 4.5. It means that the FCG rate is increased in gaseous hydrogen environment compared to air at high stress intensity ranges.

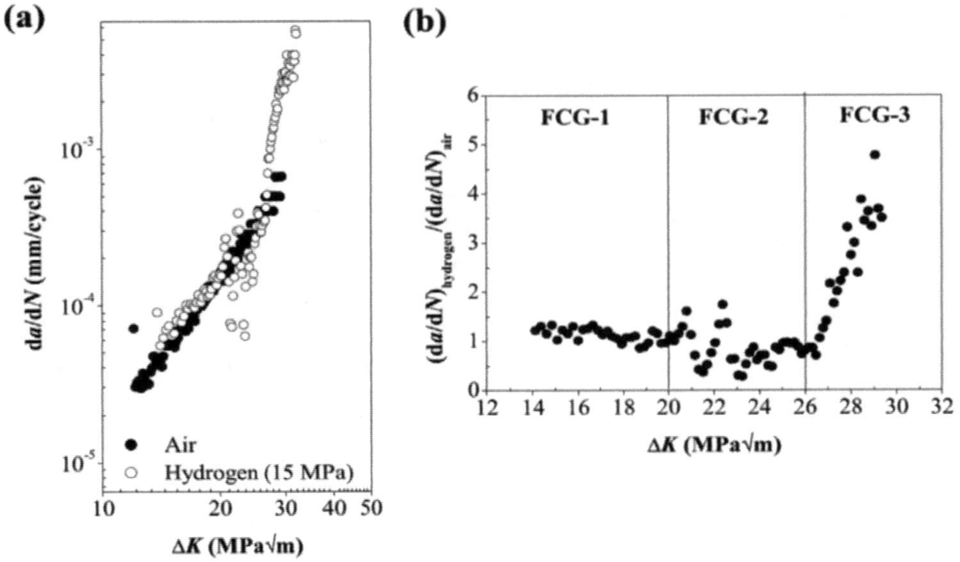

Figure 3. (a) Fatigue crack growth (FCG) rate of forged Ti-64 in air and hydrogen environment expressed in terms of d*a*/d*N* versus ΔK, (b) Fatigue crack growth rate ratio curve showing the degree of crack growth enhancement caused by the hydrogen environment.

From the test results, it can be said that the FCG rate is significantly increased in hydrogen environment at high stress intensity range and the degree of increment is dependent on the stress intensity range (ΔK). There is a critical stress intensity value ($\Delta K^* \approx 20$ MPa√m), above which the influence of hydrogen on the fatigue crack growth is clearly noticeable. Observations in the change of FCG rate in hydrogen environment are reported elsewhere for Ti-64 alloy [8, 9, 12, 13]. Pittinato et al. [8, 9] showed that FCG rate increased in hydrogen environment compared to tests in helium gas when tested at room temperature to -17°C and showed that the FCG rate is dependent on the stress intensity range. However, Evans et al. [12] and Ding et al. [13] found that the FCG rate was lower in hydrogen environment than in air at room temperature. From these studies, it can be said that the influence of hydrogen environment on the FCG rate at room temperature might be dependent on several parameters such as test conditions (cyclic frequency, applied load, stress ratio), hydrogen pressure [14] and microstructure [6].

3.2. Fractography

Figure 4(a) and figure 4(b) show an overview of the fracture surfaces observed on the samples tested in air and hydrogen environment. In addition, specific regions of interest are marked that describe the influence of hydrogen on FCG rate as shown in figure 3(b). The fracture surface in air was denoted as FCG-1. However, in hydrogen environment, the fracture surfaces are classified into three distinct regions FCG-1, FCG-2 and FCG-3 as mentioned in section 3.1 (see figure 4(d-f)). Figure 5, figure 6 and figure 7 show the high-resolution micrographs of representative fracture surfaces observed in these regions. Table 2 provides a brief summary of the fracture surface features. Analysis of the entire fracture surfaces in air and hydrogen environment at $\Delta K \leq 20$ MPa√m (i.e. FCG-1 region) showed ductile features (see figure 4(c), figure 4(d)). Here, the fracture surfaces mainly consist of fatigue striations perpendicular to the fatigue crack propagation direction, which vary in size with ΔK (see figure 5(a) and figure 5(b)). At $20 \leq \Delta K \leq 26$ MPa√m (i.e. FCG-2 region), the fracture surface in hydrogen environment consists of a mixture of flat areas containing features which look like crack arrest marks (see figure 6(a) and figure 6(e-f)), secondary cracks (see figure 4(e) and figure 6(a-f)), and fatigue striations similar to the one observed in FCG-1 (see figure 6(d)). The variation in fracture surface appearance at these intermediate ΔK values agrees with the fluctuations observed in da/dN ratio values (see figure 3(b)). Finally, at $\Delta K > 26$ MPa$\sqrt{}$m, the fracture surface in air consists of fatigue striations (see figure 7(a)). However, in hydrogen environment (i.e. in FCG-3 region), the fracture surface has a substantial increase of secondary cracks (see figure 4(f), figure 7(b)), and looks brittle.

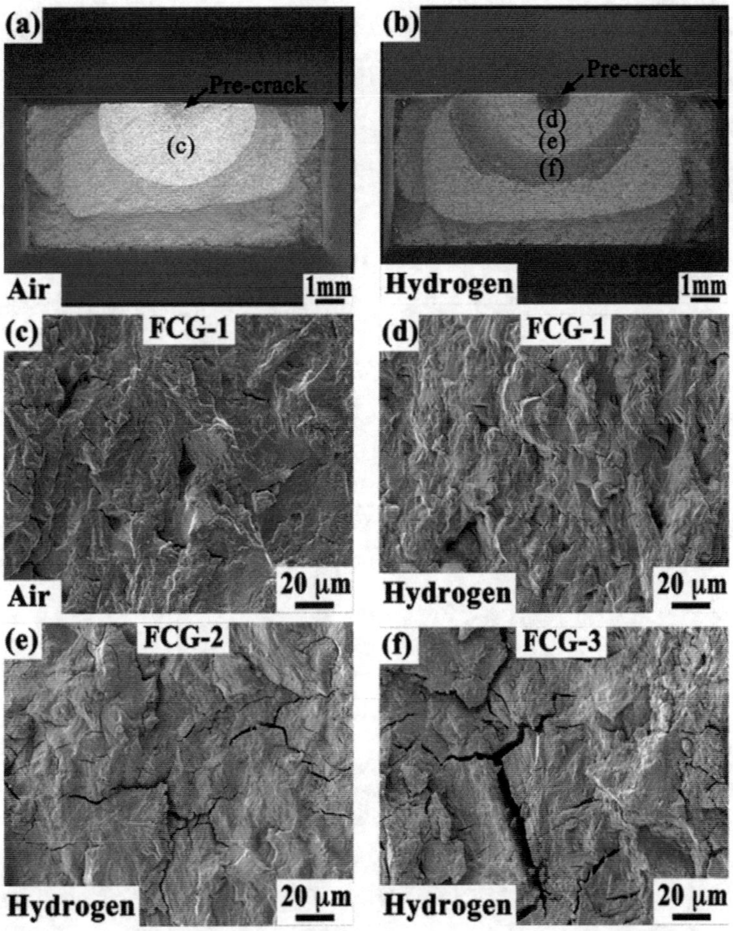

Figure 4. Micrographs showing the macroscopic fracture surfaces in (a) air and (b) hydrogen. Magnified images (c,d) show the similar appearance at low ΔK values, (e, f) show the presence of secondary cracks in hydrogen environment beyond the critical ΔK^* values. Arrows indicates the direction of crack propagation.

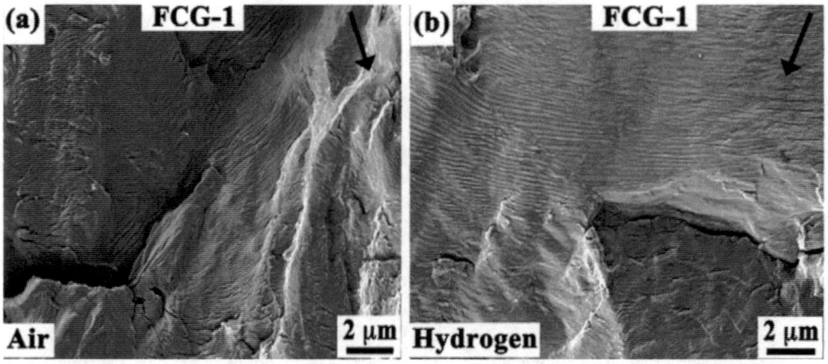

Figure 5. Representative micrographs of fatigue fracture surfaces in FCG-1 region ($\Delta K \leq 20$ MPa√m) of samples tested in air (a) and hydrogen (b) showing the presence of striations. Arrows indicate the crack propagation direction.

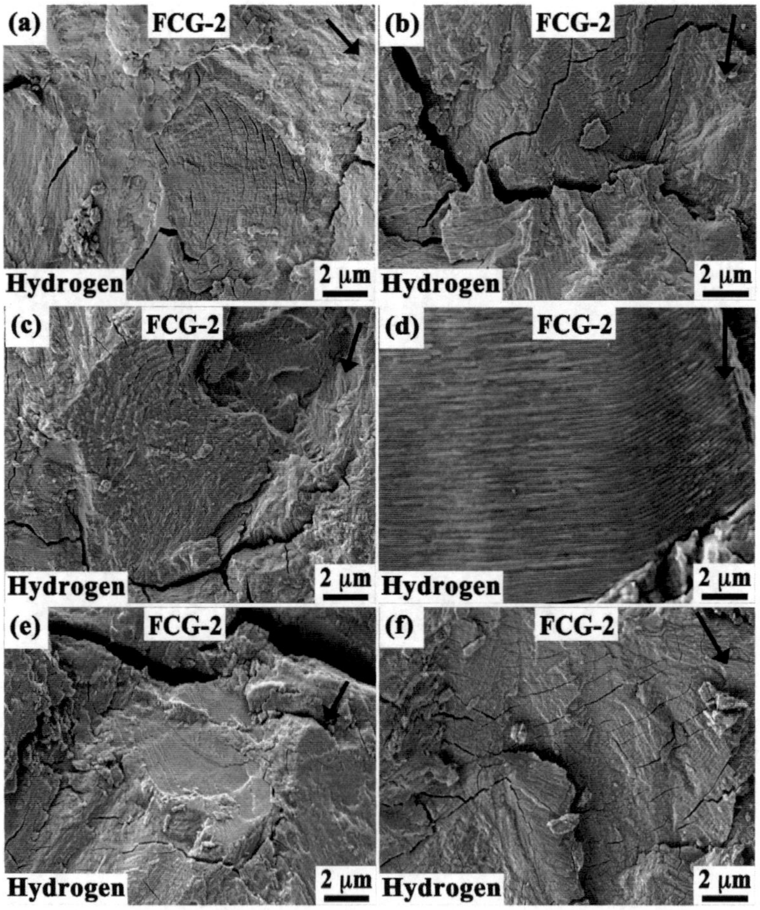

Figure 6. Representative micrographs of fatigue fracture surfaces in hydrogen environment along the FCG-2 region: (a-c) $\Delta K \approx 22$ MPa√m, (d) $\Delta K \approx 23$ MPa√m, (e) $\Delta K \approx 24$ MPa√m and (f) $\Delta K \approx 26$ MPa√m. Arrows indicate the crack propagation direction.

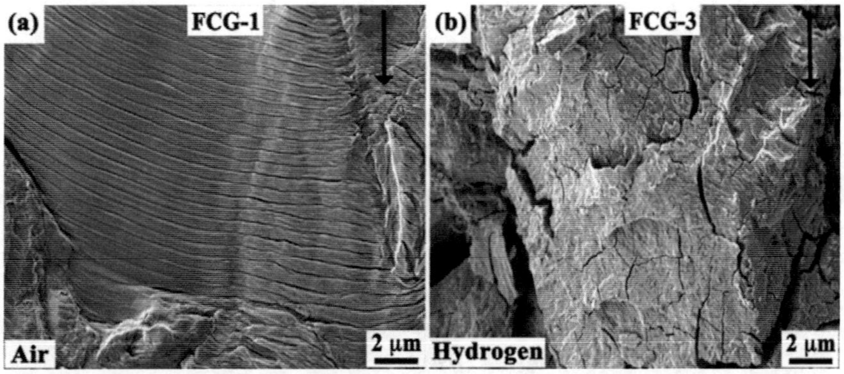

Figure 7. Representative micrographs of fatigue fracture surfaces at $\Delta K \approx 28$ MPa√m in (a) air and (b) hydrogen environment. Arrows indicate the crack propagation direction.

Table 2. Summary of the fatigue fracture surface appearance in the different FCG regions.

Region	ΔK (MPa√m)	Environment	
		Air	Hydrogen
FCG-1	≤ 20	Mostly striations	Mostly striations
FCG-2	20 – 26	Mostly striations	Mix of striated areas and flat regions. Decreasing amount of striated areas and increasing density of secondary cracks and crack arrest marks with increasing ΔK
FCG-3	> 26	Mostly striations	No striations, high density of secondary cracks

The observations made on fracture surfaces at different regions in hydrogen environment are consistent with the FCG behaviour described in section 3.1. In FCG-1 region (i.e. at lower ΔK), fracture surfaces mainly consist of fatigue striations; presumably the stress is too low for accumulation of hydrogen ahead of the crack tip. Instead the crack growth is dominated by the same mechanisms as in air. However, in FCG-2 region (i.e. at intermediate ΔK), the effect of hydrogen is clearly noted. Here, fracture surfaces mostly consist of secondary cracks that might have occurred due to an increase of stresses locally at the crack tip in presence of hydrogen and resulting in crack branching, which may lower the da/dN values [15]. In addition, the fracture surfaces also consist of relatively flat areas with feature resembling crack arrest marks that might be the result of a stress induced hydride formation and cracking as reported by Nelson et. al. [6, 7]. It is presumed that there could be enough time available for hydrogen to accumulate in front of the crack tip and precipitate as brittle hydrides at these intermediate stress intensity ranges, similar to what was observed by Shih et. al. [16] in a Ti-4Al alloy. Therefore, it seems that the fluctuation in FCG rate in hydrogen at these stress intensity levels could be attributed to the competing effects due to crack branching and brittle fracture of hydrides. Finally in FCG-3 region (i.e. at higher ΔK), the fracture surface consists of rough surface morphology with higher density of secondary cracks and a brittle appearance. At these higher stress intensities, it seems that the crack growth by fracture of hydrides is not possible, as there is no sufficient time for hydride formation. Instead the hydrogen diffused would accumulate at the crack tip and resulting in increase of the stresses locally by enhancing the mobility of dislocations. The increase of dislocations on a very local scale because of hydrogen leads to a highly localised fracture, this phenomena is commonly termed as hydrogen enhanced localised plasticity [16-18]. Thus, the increase of plastic deformation locally at the crack tip can then result in brittle fracture surface appearance, which inherently increases the FCG rate.

4. Conclusions

In the present study, fatigue crack growth resistance of forged Ti-64 was determined in air and gaseous hydrogen environment (15 MPa) at room temperature. The conclusions made from the studies are:

- FCG rate in hydrogen environment remained unaffected at stress intensity ranges below a threshold value i.e. around 20 MPa√m. However it was fluctuating at intermediate stress intensity ranges and increased drastically at higher stress intensity ranges.

- Fatigue fracture surfaces in air consisted of fatigue striations. In contrast, fracture surfaces in hydrogen environment consisted of fatigue striations below a threshold value; brittle flat surfaces and secondary cracks at intermediate stress intensities; and highly deformed surfaces with higher amount of secondary cracks at higher stress intensity ranges.

- It has been suggested that the fluctuations in the crack growth rate at intermediate stress intensity factors could be a result of competing effect of crack deceleration because of crack branching and acceleration because of hydrogen, whereas the later prevails at higher stress intensity ranges.

5. References

[1] Boyer R R 1996 *Mater. Sci. Eng.* A **213 (1-2)** 103-114
[2] Peters M, Kumpfert J, Ward C H and Leyens C 2003 *Adv. Eng. Mater.* **5(6)** 419-427
[3] Tal-Gutelmacher E and Eliezer D 2005 *J. Alloys Comp.* **404-406** 621-625
[4] Walter R J and Chandler W T 1969 *Effects of high pressure hydrogen on metals at ambient temperature* Final report R-7780-l Rocketdyne California USA
[5] Gaddam R, Åkerfeldt P, Pederson R and Antti M L 2011 *Proc. of The 12th World Conf. on Titanium, (Beijing)* **3** (Science Press) 1885-1888
[6] Nelson H G, Williams D P and Stein J E 1972 *Metall. Mater. Trans.* B **3(2)** 469-475
[7] Nelson H G 1976 *Metall. Trans.* A **7A** 621-627
[8] Pittinato G F 1969 *Trans. ASM* **62** 410-417
[9] Pittinato G F 1972 *Metall. Mater. Trans.* B **3(1)** 235-243
[10] ASTM E647-08, "Standard test method for measurement of fatigue crack growth rates", ASTM International, West Conshohocken, PA, DOI: 10.1520/E0647-08, www.astm.org.
[11] ASTM E740-03, "Standard practice for fracture testing with surface-crack tension specimens", ASTM International, West Conshohocken, PA, DOI: 10.1520/E0740-03, www.astm.org.
[12] Evans W, Bache M R, McElhone M and Grabowski L 1997 *Int. J. Fatigue* **19(93)** 177-182
[13] Ding Y S, Tsaya L W and Chen C 2009 *Corros. Sci.* **51** 1413-1419
[14] Nelson H G and Williams D P 1973 *Metall. Mater. Trans.* B **4(1)** 364-367
[15] Janssen M, Zuidema J and Wanhill R 2004 *Fracture Mechanics*, 2nd edition (London and Newyork: Spon Press: Taylor & Francis group) 319-320
[16] Shih, D S, Robertson I M and Birnbaum H K 1988 *Acta Mater.* **36(1)** 111-124
[17] Robertson I M 2001 *Eng. Fract. Mech.* **68(6)** 671-692
[18] Birnbaum H K and Sofronis P 1994 *Mater. Sci. Eng.* A **176(1-2)** 191-202

Acknowledgments
The authors kindly acknowledge Graduate School of Space Technology and National Aviation Engineering Research Programme (NFFP) for funding the research work. We would like to acknowledge Dr. Johanne Mouzon for providing the high resolution scanning electron microscopy facilities at Department of Civil, Environmental and Natural Resources Engineering, Division of Sustainable Process Engineering, Luleå University of Technology, Sweden.

IOP Conf. Series: Materials Science and Engineering **48** (2013) 012011 doi:10.1088/1757-899X/48/1/012011

The effect of crystallographic orientation on solid metal induced embrittlement of Ti-8Al-1Mo-1V in contact with copper

P Åkerfeldt[1]*, R Pederson[1,2] , M-L Antti[1], Y Yao[3] and U Klement[3]
[1]Division of Materials Science, Luleå University of Technology, SE 97181 Luleå, Sweden
[2]Research & Technology Centre, GKN Aerospace Engine Systems, SE 46181 Trollhättan, Sweden
[3]Department of Materials and Manufacturing Technology, Chalmers University of Technology, SE 41296 Göteborg, Sweden

*Corresponding author: pia.akerfeldt@ltu.se

Abstract. Solid metal induced embrittlement (SMIE) occurs when a metal experiences tensile stress and is in contact with another metal with lower melting temperature. SMIE is believed to be a combined action of surface self-diffusion of the embrittling species to the crack tip and adsorption of the embrittling species at the crack tip, which weakens the crack tip region. In the present study, both SMIE of the near alpha alloy Ti-8Al-1Mo-1V in contact with copper and its influence by crystallographic orientation have been studied. U-bend specimens coated with copper were heat treated at 480°C for 8 hours. One of the cracks was examined in detail using electron backscatter diffraction technique. A preferable crack path was found along high angle grain boundaries with grains oriented close to [0001] in the crack direction; this indicates that there is a connection between the SMIE crack characteristics and the crystallographic orientation.

1. Introduction

Metal induced embrittlement (MIE) can occur when a metal experiences tensile stress and is in contact with a metal of lower melting temperature. What distinguishes MIE is that normally ductile metals become brittle in contact with metals under certain conditions [1] and that the degradation is often undetected until the catastrophic failure occurs [2]. There are two types of MIE: embrittlement that occurs above the melting temperature of the embrittling metal (T_m), referred to as liquid metal embrittlement, and embrittlement that occurs below T_m, known as solid metal induced embrittlement (SMIE) [1,3,4]. SMIE has been investigated in the present study.

The essential condition for SMIE to occur is intimate contact between the embrittling metal and the metal substrate [2,4-8]. In order to promote SMIE, any oxides present must first be removed either chemically or mechanically. The service temperature (T) and the amount of tensile stress present are also influencing parameters; below certain threshold values, SMIE is not expected to take place [1,5,9,10]. The crack propagation rate of SMIE increases with increasing temperature, and is highest just below T_m [4,6,9]. However, the severity of SMIE is also dependent on the distance between the source of the embrittling species and the crack tip. If embrittling species are present as inclusions the crack propagation rate has been found to be high [4,7], whilst severity is expected to decrease with

IOP Conf. Series: Materials Science and Engineering **48** (2013) 012011 doi:10.1088/1757-899X/48/1/012011

increasing crack length when embrittling species are present as thin films on the surface [7]. The mechanism of SMIE is believed to be a combination of surface self-diffusion of the embrittling species to the crack tip [3,7,11,12] and adsorption of the embrittling species at crack tip [4,5,7,11]. It is generally agreed that adsorption of the embrittling metal leads to a weakening of the crack tip, but exactly how adsorption affects the crack propagation is not fully understood. Two principal theories are: (1) chemisorption reduces the stress required for tensile separation of the atoms at the crack tip leading to decohesion [1,13-15], and (2) chemisorption facilitates the nucleation or egress of dislocations at the crack tip [11,16-19].

SMIE of titanium alloys in contact with the following solid materials has been observed: silver, gold, copper and cadmium, see table 1. In 1965 Duttweiler et al. [20] concluded that the failure of titanium compressor discs resulted from SMIE caused by silver chloride. Additional studies showed that Ti-5Al-2.5Sn, Ti-8Al-1Mo-1V and Ti-7Al-4Mo alloys all exhibited SMIE when in contact with solid silver. Embrittlement caused by solid silver has also been studied by Stoltz and Stulen [21], who observed SMIE of Ti-6Al-6V-2Sn in contact with solid silver, gold and cadmium. Meyn [22], and Fager and Spurr [10] have reported SMIE caused by cadmium in contact with Ti-8Al-1Mo-1V, Ti-6Al-4V and Ti-3Al-14V-11Cr. Most recently however, the influence of solid copper has been studied; Liu [23] reported in 2006 that solid copper caused SMIE of Ti-6.5Al-3.5Mo-1.5Zr-0.3Si. The present work is a continuation of a previous study of SMIE in Ti-6Al-2Sn-4Zr-2Mo and Ti-8Al-1Mo-1V in contact with copper and gold [24]. Although SMIE has been observed in previous studies, crack propagation behaviour in relation to crystallographic orientation has not been evaluated; this is the aim of the present study.

Table 1. Previous investigations of titanium alloys and embrittling species.

Alloy	Solid embrittling species
Ti-5Al-2.5Sn	Ag [20]
Ti-8Al-1Mo-1V	Ag [20], Au [24] Cd [22], Cu [24]
Ti-6Al-4V	Cd [22]
Ti-6Al-6V-2Sn	Ag [21], Au [21], Cd [21]
Ti-6Al-2Sn-4Zr-2Mo	Au [24], Cu [24]
Ti-6.5Al-3.5Mo-1.5Zr-0.3Si	Cu [23]
Ti-7Al-4Mo	Ag [20]
Ti-3Al-14V-11Cr	Cd [22]

2. Materials and Method

A U-bend test was used to evaluate SMIE of Ti-8Al-1Mo-1V in contact with copper. The sheet material, condition and composition were in accordance with AMS4916J [25] and were statically loaded by U-bending, and then coated with 0.05-0.1 μm solid copper using sputtering technique. Specimens were subsequently heat treated at 480°C for 8 hours; details of the test method can be found in previous work [24]. Principal metallographic examination was performed in a light optical microscope at 500x magnification. The outer surface of the cross-section of the U-bend shape was examined for cracking and the locations of the observed cracks noted.

Specimens exhibiting cracking were further examined in a scanning electron microscope (SEM, LEO 1550 Gemini FEG-SEM). The cracks were studied by using backscattered electron imaging (BSE) with high contrast, which allowed the alpha and beta phases to be discriminated by compositional contrast without etching. Electron backscatter diffraction (EBSD), a characterisation technique employed in the SEM, was used to determine the local crystallographic orientation and the texture of the area of interest along the crack propagation path in a specimen. For EBSD mapping an HKL Channel-5 EBSD system with a Nordlys II detector was used. Two orientation maps were

IOP Conf. Series: Materials Science and Engineering **48** (2013) 012011 doi:10.1088/1757-899X/48/1/012011

acquired at 5000x magnification with an accelerating voltage of 20 kV and a step size of 200 nm. An average noise reduction of four neighbours was applied.

3. Results

In figure 1 a), the BSE image of a typical crack path is shown. Alpha and beta phase can clearly be discriminated by the compositional contrast as grey and white, respectively. For a better understanding of the crack propagation characteristics, EBSD measurements were performed along the crack path. To the right in figure 1, band contrast images are shown obtained at area A and B, which clearly shows that the crack is mainly propagating along the grain boundaries.

Two orientation maps were acquired along the crack path and in the respective inverse pole figures in Y-direction, the orientations of the grains along the crack were found to be close to [0001]. In figure 2 a), the orientation maps given in inverse pole figure scheme are stitched together to create a map including the whole crack path. According to the inverse pole figure colour scheme given in figure 3, the grains with the orientation close to [0001] are coloured red. In the phase map (figure 2 b)), the

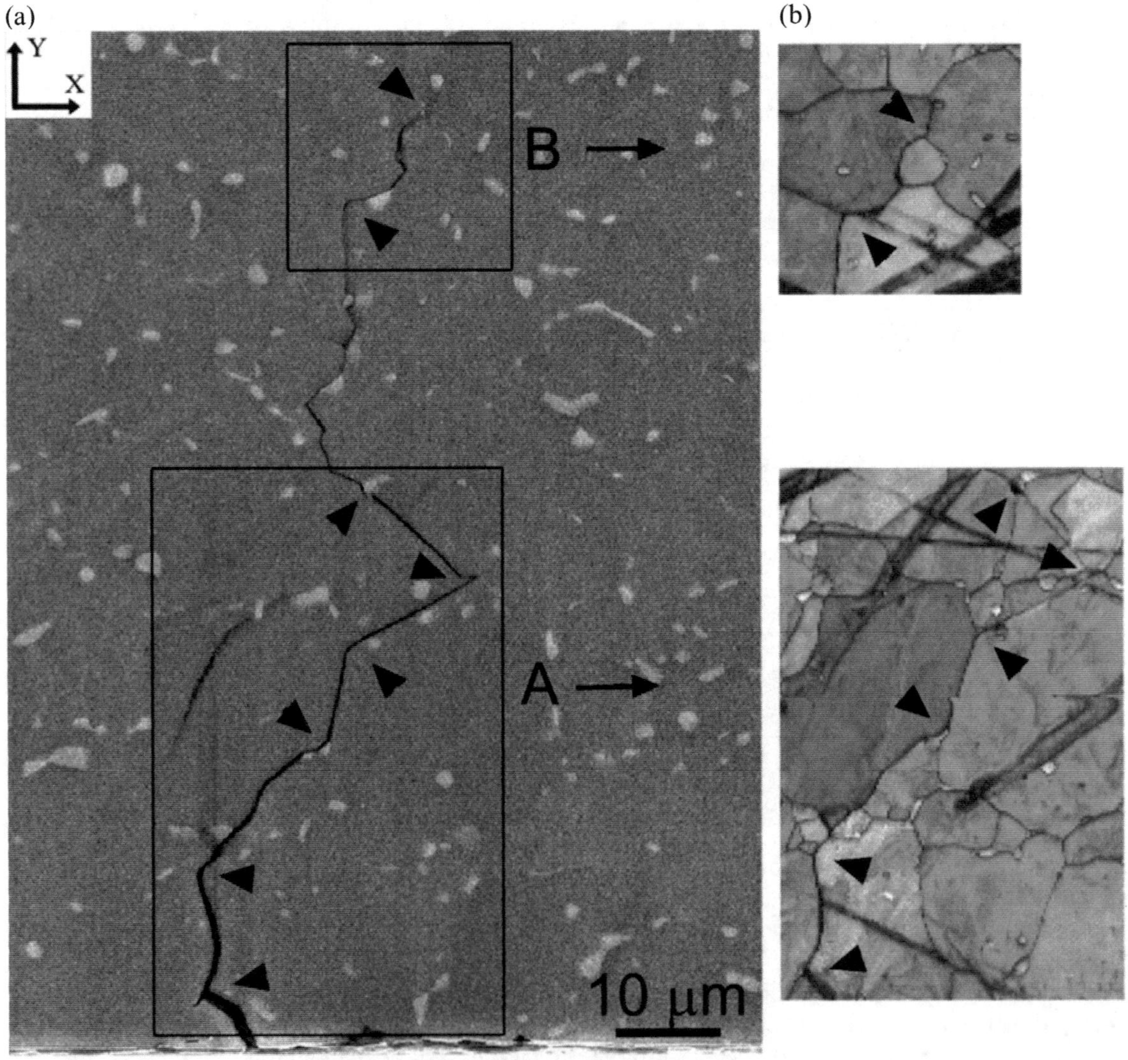

Figure 1. (a) The typical appearance of the SMIE of Ti-8Al-1Mo-1V in contact with copper and (b) the band contrast images of area A and B indicating how the crack is propagating along the grain boundaries.

alpha and the beta phases are coloured blue and red, respectively, and the orientations of the grains along the crack path represented by the unit cell. The grain boundaries in the orientation map in figure 2 a) are defined by the degree of misorientation (θ) between two neighbouring pixels. The high-angle grain boundaries (θ>10°) are given as black lines while the low-angle boundaries (3°<θ<10°) by white lines. The EBSD results indicate that the crack path is following high-angle grain boundaries and there is a tendency of the crack to propagate along [0001]-oriented grains (along Y-direction).

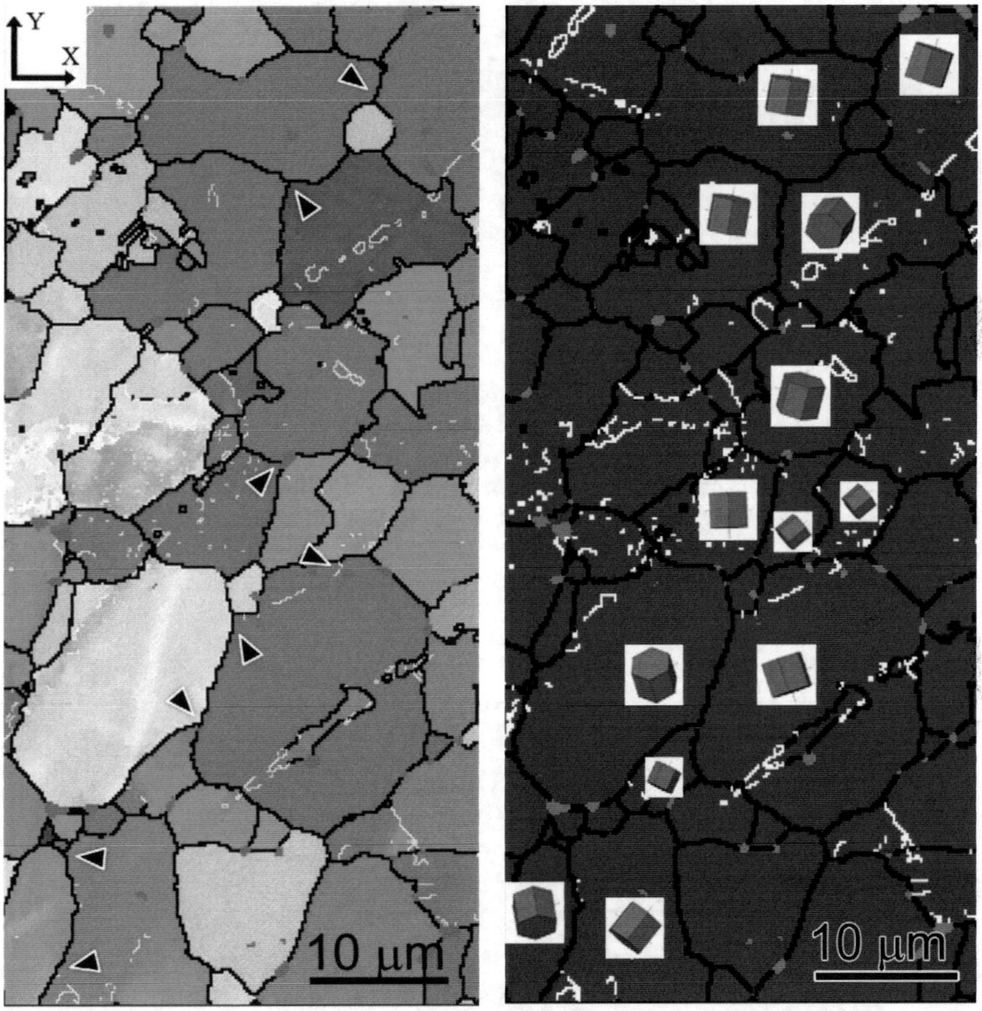

(a) **(b)**

Figure 2. (a) Orientation map in inverse pole figure colour code and (b) phase map with grain orientations along the crack path represented by their respective unit cells.

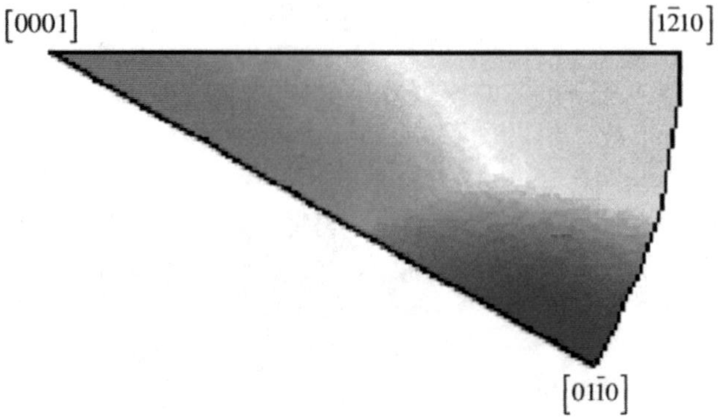

Figure 3. Inverse pole figure colour scheme used for the orientation map in figure 2.

4. Discussion

Metals with hcp (hexagonal close-packed) crystallographic structure such as alpha titanium (see grey phase in figure 1), normally fracture by cleavage fracture [26]. In this study the crack growth path was found to be intergranular, which itself is a clear indication that there is an influence of an embrittling environment. The crack propagates along grain boundaries, perpendicular to the surface and loading direction, i.e. in a direction towards high stress levels. The embrittling environment in the present study was solid copper. During heat treatment in air at 480°C however, the copper layer oxidises and consequently the embrittling species closest to the titanium surface are in the form of CuO, Cu_2O, and Cu, see figure 4. Because of the strong reducing properties of titanium, the copper oxide closest to the titanium surface is probably reduced to form titanium oxide and copper:

$$2Cu_2O(s) + Ti(s) \rightleftharpoons 4Cu(s) + TiO_2(s) \tag{1}$$

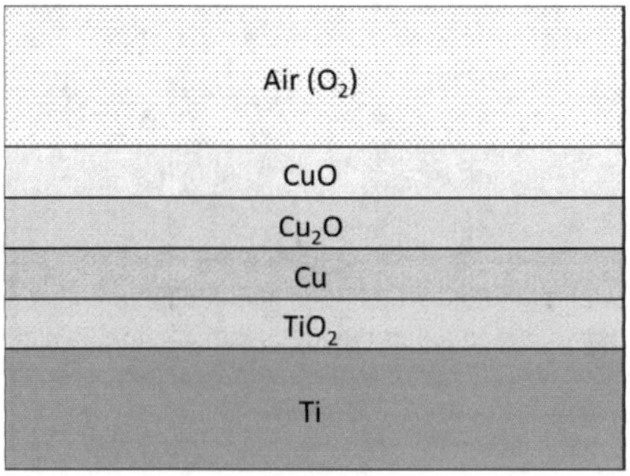

Figure 4. The assumed surface composition after heat treatment in air.

Once the crack is initiated, the embrittling species are believed to be transported by surface self-diffusion to the crack tip and adsorption of the embrittling species at the crack tip weakens the crack tip region. In previous work the characteristic self-diffusion length of copper for 8 hours at 480°C, has been calculated to be 2300 μm [27]. Hence, the diffusion length is longer than the crack length observed in the experimental study. When comparing the characteristic diffusion length with the experimental crack length however, one should also consider the incubation time and the fact that the experimental crack is following a serrated path making the crack much longer than what it first appears as. In addition the amount of copper available for surface self-diffusion needs to be taken into account since the presence of copper at the crack tip is crucial for crack growth to occur. The relationship between crack length and amount of copper available can be estimated using the following simple stoichiometric calculation. Assume that the crack is two-dimensional and that the copper atoms originate from a layer of copper atoms 10 μm wide (*w*) and 0.05 μm thick (*t*) at the surface. Moreover, assume that the crack wall is covered with at least five atom layers of copper, this to sustain the SMIE crack propagation rate. Thus, at least ten atom layers (*N*) are required for self-diffusion of copper along our two-dimensional crack. The estimated crack length, L_{crack}, can then be calculated:

$$L_{crack} = \frac{wt}{Nd} = \frac{10 \cdot 0{,}05}{10 \cdot 0{,}000256} \approx 200 \mu m \tag{2}$$

where the diameter of a copper atom (*d*) being 0.000256 μm. Thus, with an estimated thickness of sputtered copper layer of 0.05 – 0.1 μm, the calculated stoichiometric crack length lies between 200 and 400 μm. Taking the diffusion rate and availability of copper atoms into consideration, the transport mechanism of surface self-diffusion appears to be applicable for the present results.

The mechanism of adsorption at the crack tip is complex. It is generally agreed that chemisorption of the embrittling species weakens the crack tip region, but exactly how is debatable [1,17]. The result of this EBSD study shows that the crack propagates principally along high-angle grain boundaries, which is in agreement with theories of adsorption: the thermodynamic driving force for adsorption is the reduction in surface free energy, which is larger for high-angle grain boundaries [28]. However, when the orientation of grains along the crack path is considered, a favourable crack growth is observed along grains oriented close to [0001] in Y-direction (see figure 2). A possible reason may be the coordination number of the bulk atoms, which influences surface energy. Along the crack there is a layer of copper and copper oxides, see figure 5, and at the interface of the crack tip both oxygen and copper atoms bond to the titanium lattice. In the present study, several grains along the crack are oriented with $(01\bar{1}0)$ parallel to the local crack path direction. When comparing the coordination number of bulk atoms in (0001) with $(01\bar{1}0)$, the surface free energy of (0001) is lower because it is more densely packed. Hence, adsorption of copper is more likely to occur on $(01\bar{1}0)$, since the driving force (potential reduction in energy) is higher. In addition, oxygen atoms are attracted to octahedral interstices in the hcp crystallographic structure, which are available in the crystallographic structure oriented $(01\bar{1}0)$. Thus by considering the atomic configuration in $(01\bar{1}0)$, adsorption of both copper and oxygen atoms is more likely to occur in that plane than in others such as (0001).

IOP Conf. Series: Materials Science and Engineering **48** (2013) 012011 doi:10.1088/1757-899X/48/1/012011

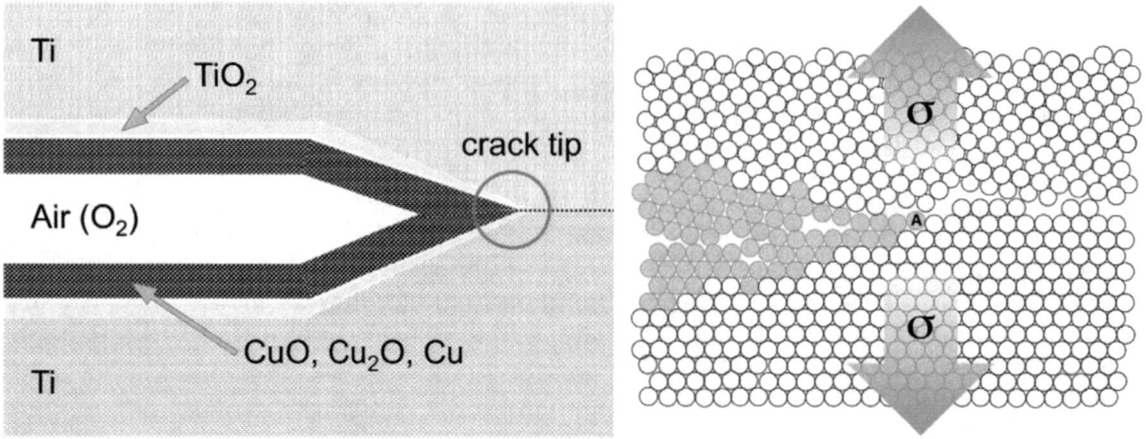

(a) **(b)**

Figure 5. (a) Schematic illustration of the crack wall and (b) the crack tip. Copper (and oxygen) enters the crack by surface self-diffusion and at the crack tip the copper atom "A" is adsorbed. The crack growth occurs under the tensile stress σ.

Another contributing reason to favourable crack growth along grains oriented close to [0001] in the crack direction (Y) could originate in the crystallography of titanium (alpha) and copper having hcp and fcc (face-centred cubic) structure, respectively. Studies of grain boundary segregation report that some grain boundaries are more prone to the segregation than others. Partly that is explained by the orientation relationship, that grain boundaries with preferable crystallography in relation to the segregate may be able to accommodate more segregates [29]. However, the work carried out in this area is limited, and no previous work has been done in this respect regard to SMIE of titanium. When studying hcp and fcc systems, four principal orientation relationships can be predicted [30], see figure 6, corresponding to the directions and planes in the hcp unit cell as given in figure 7.

With regard to SMIE, the orientation relationship could influence the crack path by attracting copper atoms to sites where they can adopt a structure similar to fcc. In figure 2 b), the crack path can be found along grains oriented in the directions $[10\bar{1}0]$ or $[11\bar{2}0]$ relative to the crack path, with one exception in the middle of the orientation map where the crack has propagated through a low angle boundary and across another grain, both oriented with (0001) parallel to the crack direction. However, the tendency of a favourable crack growth along grains orientated in the directions $[10\bar{1}0]$ and $[11\bar{2}0]$ relative to the crack path corresponds well with the hcp directions listed by the orientation relationships in figure 6. For instance, when $[10\bar{1}0]$ is parallel to [112] and (0002) is aligned with $(\bar{1}11)$ there is an orientation relationship between the hcp and the fcc crystallographic structure. Accordingly, for that specific crystallographic orientation, the atomic misfit between hcp and fcc is low, which enables the copper atoms to adopt a structure similar to fcc. Hence, as with segregation, a certain grain boundary orientation could attract and accommodate more copper, than another resulting in a heterogeneous distribution of copper atoms and thereby the crack growth might be preferable in certain directions.

IOP Conf. Series: Materials Science and Engineering **48** (2013) 012011 doi:10.1088/1757-899X/48/1/012011

A: $\left[11\bar{2}0\right]_{hcp} \| [110]_{fcc}$, $(0002)_{hcp} / (\bar{1}11)_{fcc}$

B: $\left[11\bar{2}0\right]_{hcp} \| [110]_{fcc}$, $(1\bar{1}01)_{hcp} / (\bar{1}11)_{fcc}$

C: $\left[10\bar{1}0\right]_{hcp} \| [112]_{fcc}$, $(1\bar{2}10)_{hcp} / (\bar{2}20)_{fcc}$

D: $\left[10\bar{1}0\right]_{hcp} \| [112]_{fcc}$, $(0002)_{hcp} / (\bar{1}11)_{fcc}$

Figure 6: Four predicted orientation relationships between the hcp (alpha titanium) and fcc (copper) systems.

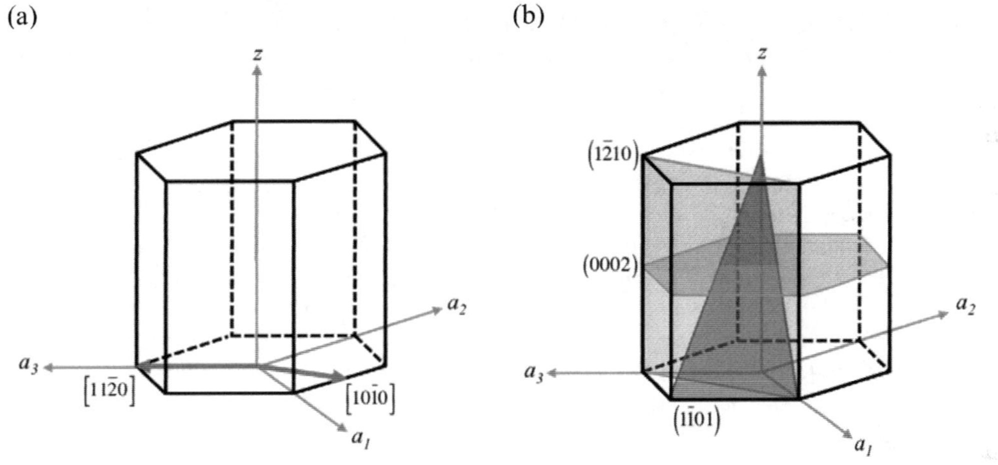

Figure 7. Orientation relationships in the hcp unit cell: (a) directions and (b) planes.

However, it is important to realise the complexity of adsorption; there are several theories that describe how the embrittling species weakens the crack tip region. The influence of adsorption on the crack tip will depend on the strength of adsorption, which in turn is dependent on several parameters, such as the surface energy and the solubility of the embrittling species in the matrix. Moreover, one should also keep in mind that the atomic bonds at the crack tip in the vicinity of atom "A" in figure 5 b), are strained by the tensile stress σ. Therefore, when the crack advances, crack growth is a result of both local stress and weakening of atomic bonds because of adsorption; depending on their magnitudes, the crack tip may experience either decohesion or dislocation emission.

5. Conclusions

The results of the present study agree with theories of surface self-diffusion and adsorption. Based on the observations of crack propagation path in relation to crystallographic orientation, the following conclusions can be drawn:

- Cracking in SMIE propagates principally along high-angle boundaries.
- Cracks propagate preferably along [0001] oriented grains in the crack direction, indicating that additional parameters influence the SMIE crack characteristics.

Acknowledgements
The authors would like to acknowledge Dr Dennis Lundström in the department of Materials Technology at GKN Aerospace Engine Systems for his practical assistance in the project, GKN Aerospace Engine Systems for the close collaboration and the graduate school in Space Technology for encouraging the project. Financial support from NFFP (the National Aviation Engineering Research Programme) is also highly appreciated.

References
[1] Kamdar M H 1983 Liquid Metal Embrittlement *Embrittlement of Engineering Alloys (Treatise on Material Science and Technology* vol 25) ed C L Briant and S K Banerji (London Academic Press) chapter 9 pp 362-455
[2] Lynch S P 1994 Failures of structures and components by environmentally assisted cracking *Eng. Failure Anal.* **1** 77-90
[3] Gordon P 1978 Metal induced embrittlement of metals – an evaluation of embrittler transport mechanism *Metallurgical Transactions A* **9** 267-73
[4] Lynch S P 1992 Metal-induced embrittlement of materials *Materials Characterization* **28** 279-89
[5] Kamdar M H 1985 Solid metal induced embrittlement of metals *Proc. of 6th International Conference on Fracture* (New Delhi) pp 3837-49
[6] Kamdar M H 1988 Solid metal induced embrittlement *J. Electrochem. Soc.* **37** 53-9
[7] Lynch S P 2008 Failures of structures and components by metal-induced embrittlement *Journal of Failure Analysis and Prevention* **8** 259-74
[8] Lynch S P 2003 Failures of engineering components due to environmentally assisted cracking *Practical Failure Analysis* **3** 33-42
[9] Kolman D G 2003 Solid metal induced embrittlement *ASM Handbook* ed S D Cramer and B S Convino (ASM International)
[10] Fager D N and Spurr W F 1970 Solid cadmium embrittlement: titanium alloys *Corrosion* **26** 409-19
[11] Lynch S P 1989 Solid-metal-induced embrittlement of aluminium alloys and other materials *Mater. Sci. Eng. A* **108** 203-12
[12] Lynn J C, Warke W R and Gordon P 1975 Solid metal-induced embrittlement of steel *Mater. Sci. Eng. A* **18** 51-62
[13] Westwood A R C and Kamdar M H 1963 Concerning liquid metal embrittlement, particularly of zinc monocrystals by mercury *The Philosophical Magazine* **8** 787-804
[14] Stoloff N S and Johnston T L 1963 Crack propagation in a liquid environment *Acta Metallurg.* **11** 251-6
[15] Westwood A R C, Preece C M and Kamdar M H 1967 Application of crack propagation criterion to liquid-metal embrittlement; cleavage of aluminium monocrystals in liquid gallium *Trans. Am. Soc. Metals* **60** 723-5
[16] Lynch S P 1981 Liquid-metal embrittlement in an Al 6%Zn 3%Mg alloy *Acta Materialia* **29** 324-40
[17] Lynch S P 1984 Metal-induced embrittlement of ductile materials and dislocation emission from crack tips *Scripta Metallurgica* **18** 509-13
[18] Lynch S P 1979 Mechanisms of fatigue and environmentally assisted fatigue *Fatigue Mechanisms* ed J T Fong (American society for testing and materials) pp 175
[19] Lynch S P 1978 Effect of environment on fracture – mechanisms of liquid-metal embrittlement, stress-corrosion cracking and corrosion-fatigue *Advances in Research on the Strength and Fracture of Materials* (Waterloo)

[20] Duttweiler R E, Wagner R R and Antony K C 1965 An investigation of stress-corrosion failures in titanium compressor components *Stress-Corrosion Cracking of Titanium: A Symposium Presented at the Fifth Pacific Area National Meeting* (Washington: ASTM) pp 152-78
[21] Stoltz R E and Stulen R H 1979 Solid metal embrittlement of Ti-6Al-6V-2Sn by cadmium, silver and gold *Corrosion* **35** 165-9
[22] Meyn D A 1973 Solid cadmium cracking of titanium alloys *Corrosion* **29** 192-6
[23] Liu D –X, Liu S –M and Fan G –F 2006 Investigation on behavior of solid copper-induced embrittlement of titanium alloy *J. Aeronaut. Mater.* **26** 1-5
[24] Åkerfeldt P, Pederson R and Antti M–L 2011 Solid metal induced embrittlement of titanium alloys in contact with copper *Proc. of The 12th World Conference on Titanium (Ti-2011)* (Beijing) pp 1868-71
[25] AMS 4916J 2006 Aerospace Material Specification: Titanium alloy sheet, strip, and plate 8Al-1Mo-1V duplex annealed (Warrendale: SAE International)
[26] Joshi V A 2006 Titanium Alloys – an Atlas of Structures and Fracture Features (Boca Raton, FL: Taylor and Francis Group)
[27] Åkerfeldt P, Pederson R and Antti M–L 2011 Investigation of the influence of copper welding electrodes on Ti-8Al-1Mo-1V and Ti-6Al-2Sn-4Zr-2Mo with respect to solid metal induced embrittlement *IOP conference series: Material Science and Engineering vol 31 (2012)*
[28] Shewmon P G 1998 Grain boundary cracking *Metallurgical and Materials Transactions: A* **29A** pp 1535-44
[29] Briant C L 2001 Grain boundary structure, chemistry, and failure *Materials Science and Technology* **17** 1317-23
[30] Zhang S, Chen H, Ren H and Kelly P M 2008 Crystallography of the simple HCP/FCC system *Metallurgical and Materials Transactions: A* **39A** 1077-86

IOP Conf. Series: Materials Science and Engineering **48** (2013) 012012 doi:10.1088/1757-899X/48/1/012012

Case study of grate-chain degradation in a Grate-Kiln process

Erik A A Nilsson[1], L Pettersson[2] and M-L Antti[1]
Luleå University of Technology, 971 87 Luleå, Sweden
[2]LKAB, 981 86 Kiruna, Sweden

erik.1.nilsson@ltu.se

Abstract. Austenitic stainless steels are often used in high temperature applications due to their resistance to corrosion. Grate-Kiln processes that sinter iron ore pellets use grate-chains which are made of austenitic stainless steel to withstand the severe environment. It has been shown, however, that the grate-chain is affected by several degrading mechanisms in the harsh environment of the sintering process. A grate-chain that has been in service for 8 months was investigated in order to find the mechanisms of degradation.

Results show that slag products are accumulated on the grate-chain and interact with the steel as hot corrosion. The stainless steel is believed to be sensitized against inter-granular attack by carburization followed by inter-granular attack. The resistance towards degradation seems to decrease with time which is suggested to be caused by depletion of chromium.

1. Introduction

Austenitic stainless steels are regularly used in applications for high temperatures and demanding atmospheres due to their chromium oxide forming capabilities and the thermal stability of austenite [1]. An application for these steels is in the Grate-Kiln process which pelletizes iron ore pellets. The pelletizing process, or induration can be seen in figure 1. The grate-chain (A) is made of austenitic stainless steel and transports the green pellets to the rotary kiln indurator (furnace) (B) where the pellets will be sintered. During the transportation the wet green pellets are dried and pre-heated by heated air. Drying and pre-heating of the pellets are performed in four different stages (C-F) with different gas temperatures flowing through slits in the grate-chain and through the bed of pellets. The main problem when investigating and evaluating degradation in industry processes is the irregularity of the environmental properties during the operation. Different quantities of additives in the pellets make both the temperature, type of particles and chemical composition of the gas fluctuate.

The grate-chain consists of several grate-links that are linked together which can be seen in the schematic image of the Grate-Kiln indurator in figure 1. The grate-link has several slits and ribs which allows the gas to flow through the grate-link and heat the pellets. The temperature of the heated air ranges between 200 and 1200°C.

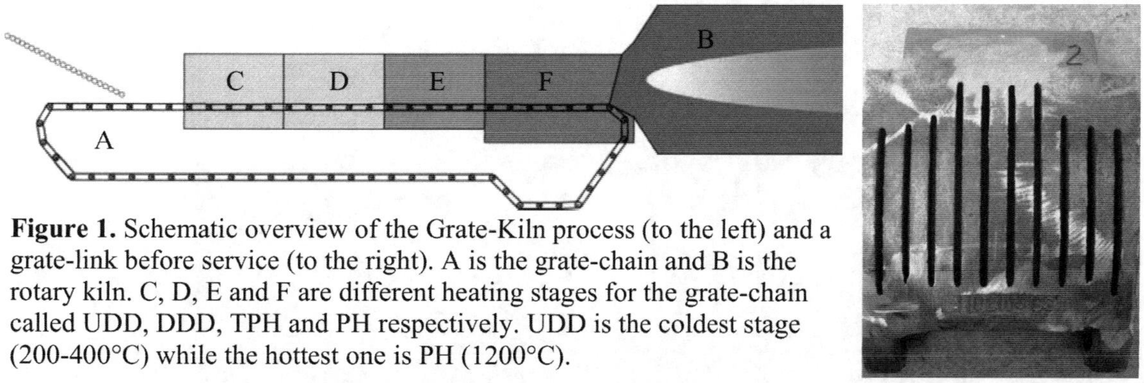

Figure 1. Schematic overview of the Grate-Kiln process (to the left) and a grate-link before service (to the right). A is the grate-chain and B is the rotary kiln. C, D, E and F are different heating stages for the grate-chain called UDD, DDD, TPH and PH respectively. UDD is the coldest stage (200-400°C) while the hottest one is PH (1200°C).

Process case studies have mostly been focusing on the petroleum industry [2-4] or heat exchange applications [5]. To the authors knowledge there are few investigations concerning the degradation of stainless steels in pellet sintering processes. Bani P. Mohanty et al. investigated hot corrosion of grate-bars within taconite indurators [6]. They concluded that degradation of stainless steel grate-bars by sulfuric salts of alkali metals was more severe when mixed together with chlorides of the same alkali metals. The alkali metals were vaporized from the pellets and later condensated as salts onto the grate-bars due to their lower temperature. This mechanism was repeated during each heating cycle and increased the concentration of the alkali salts. It was also stated that the resulting oxides from hot corrosion were more prone to spallation by thermal cycling and thereby significantly reduced the life-time of the grate-bars [7]. J-Y Liu et al. [8] studied the addition of boron, zirconia and manganese to a stainless steel in order to enhance the life-time of grate-grids in a Grate-Kiln. The results show that the life-time was enhanced from 8 months without the additions to 18 month with the additions. However, their study mostly focused on high temperature corrosion without the influence of alkali metals. These different studies show that it is necessary to investigate each environment separately in order to clarify which mechanism that governs the degradation of the material. The present study investigates the degradation mechanism of a grate-link that has been serving in an indurator during 8 months.

2. Material and Method

2.1. Material

The grate-link that was investigated had been in service for 8 months in a Grate-Kiln indurator from Metso Minerals which is operated by the Swedish mining company LKAB. The grate-link is presented in Figure 2. The material of the grate-link is an austenitic castable stainless steel named HH in the ACI standard or A 297 HH in the ASTM standard. The chemical composition of the material as stated by LKAB can be seen in table 1, together with the standard specification for ACI HH. The HH material is the most common austenitic steel for high temperature applications as it has a good resistance to corrosion.

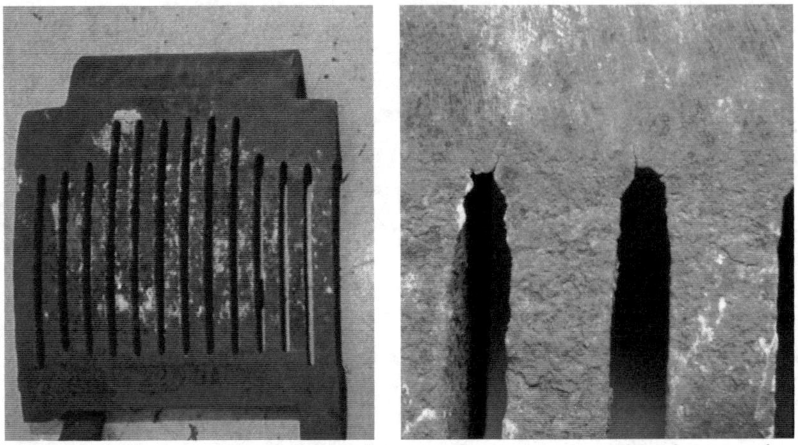

Figure 2. A grate-link after being in use for 8 months. The photo to the right shows cracks at the edges of the slits.

Table 1. Chemical composition of the material in the grate-chain and standard specifications for ACI HH.

Alloying element	C	Si	Mn	Cr	Ni	Mo	Fe
Material	0,272	0,94	0,55	25,53	10,83	0,24	Bal.
ACI HH (ASTM A297)	0,2-0,5	2	0-2	24-28	11-14	0-0,5	Bal.

The chemical composition of the environmental gases inside the indurator have been evaluated by LKAB and contains other than nitrogen; 15-20% of oxygen, 2-5% of carbon dioxide, 20-100ppm of carbon monoxide, 50-100ppm of NO_x gases, 2-5ppm of sulfur dioxide and 5-10% of water vapor. An unknown small amount of vaporized Na and K is also assumed to exist in the environmental gas, originating from the coal fuel and the pellets.

2.2. Visual inspection
The grate-chain has been inspected during service by operators and engineers and their observations are taken into account in this study in order to increase the understanding of the degradation of the grate-chain.

2.3. Investigation of deposits
Thick deposits were found on the grate-links. In order to determine if these deposits originated from the material or from the process they were analyzed with Inductively Coupled Plasma – Sector Field Mass Spectroscopy (ICP-SFMS) to give a quantitative estimation of the chemical content.

2.4. Overall degradation
To evaluate the reduction of thickness of the ribs each rib was cut so that the cross section could be measured, see figure 3. Cut A was done first to cut all the ribs loose from the grate-link. The thickness of the ribs were measured and compared to casting specifications at two positions on the rib; in the middle of the rib and close to the slit-edge. The oxide thickness and chemical content of the ribs were of interest so each rib was cut in the middle to get a cross section piece. To avoid any extensive amounts of water soluble oxides to be removed during cutting the ribs were covered in a layer of

IOP Conf. Series: Materials Science and Engineering **48** (2013) 012012 doi:10.1088/1757-899X/48/1/012012

Lacquer from METACOAT®. The gradient of chemical elements at the surface of the specimen and inwards was investigated by Energy Dispersive Spectroscopy (EDS).

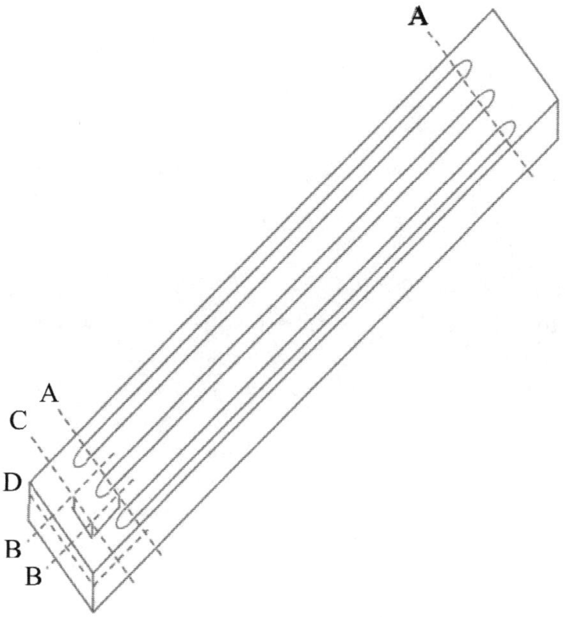

Figure 3. Cut-up scheme of a section of the grate-link. Cut A separates the ribs from the grate-link. Cut B separates each slit-edge that will be investigated. To reduce the size of the samples cut C and D are necessary.

2.5. Crack investigation at slit-edges

To evaluate the cracks in the slit-edges the grate-link was cut according to figure 3, slice B, C and D. The resulting samples were polished so that the cracks and their propagation characteristics could be evaluated by scanning electron microscopy (SEM). The cracks were investigated by EDS to see possible chromium depletion, signs of carburization and chemical content inside the cracks.

2.6. Instruments

The optical microscope used during the evaluation was a Nikon Eclipse MA 200. SEM imaging and EDS measurements was carried out by a JEOL JSM 6460LV with an Oxford INCA system. Hardness measurements were done by a Matsuzawa MXT-α Vickers micro hardness tester. The chosen load for hardness measurements was 300 gram with a loading time of 15 seconds.

3. Results

3.1. Visual inspection and deposits

Visual inspection of the grate-links during service showed that there is a clear pattern of the degradation that can be divided into three different steps.

1. During the initial period the grate-links get thicker by deposits. Since no apparent degradation is visible underneath the deposits it is concluded that the deposits do not originate from the grate-links but from the induration process.

2. After the initial period the material starts to degrade at an accelerated rate which can be seen by the visual loss of thickness. The deposits are thick and flaky and can easily be removed from the grate-link.

3. Cracks start to arise at stressed locations at the slit-edges and eventually on the ribs, perpendicular to their elongation direction.

Results from the elemental evaluation by ICP-SFMS of the deposits on the grate-link are shown in table 2. A relatively high amount of alkaline metals can be seen together with sulfur.

Table 2. Analysis of the elements in the deposits performed by ICP-SFMS.

	Si	Al	Ca	Fe	K	Mg	Mn	Na	Cr	Ni	S	C
wt%	1,98	1,53	0,68	50,0	3,98	0,35	0,08	1,06	4,85	1,26	1,98	0,10

3.2. Overall degradation

3.2.1. Relative thickness loss. The thickness of the ribs before and after a service life of 8 months was compared to establish a relative thickness loss in order to see if the grate-link has been evenly degraded. The thickness was measured at two different locations at the ribs, one measurement close to the slit-edge of the grate-link and a second measurement on the middle of the rib. The relative thickness loss can be seen in figure 4. It is clear that the ribs have not been evenly degraded. The center rib (number 6) has the highest loss of relative thickness while ribs 1 and 11 (close to the sides of the grate-link) have the smallest loss of thickness. There is a large difference between the measurements close to the slit-edge of the grate-plate and the middle of the ribs for all positions. The middle part of the ribs suffers a significantly larger degradation than those closer to the slit-edge of the grate-link. Ribs 1 and 11 have an increased thickness due to deposition.

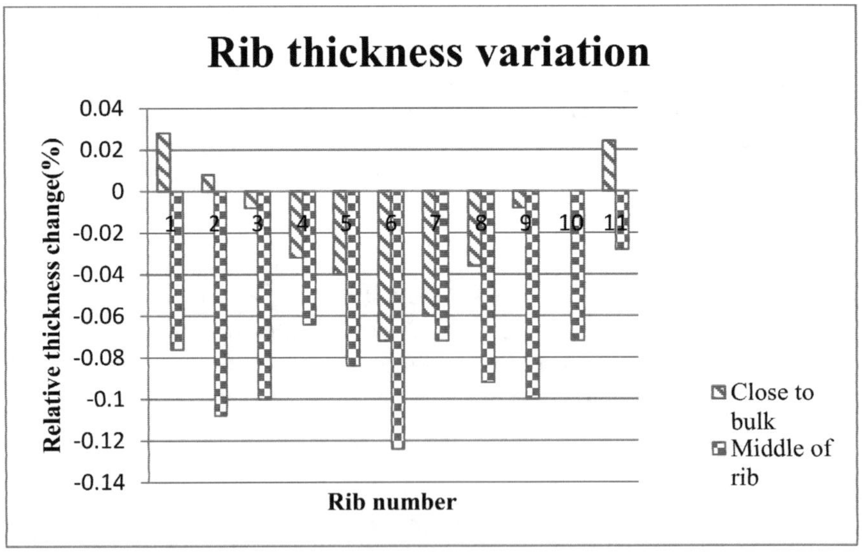

Figure 4. Rib thickness variation with the position of the rib. Rib 1 and 11 are furthest from the center of the grate-chain, while rib 6 is in the center.

3.2.2. Oxide layer investigation of rib cross section. Figure 5 shows back-scattered electron (BSE) images of the cross section of the oxide on the surface of the metal. Depositions on the metal surface and an underlying oxide layer can be seen in figure 5 a). Image b) in figure 5 shows thick grain boundaries beneath a cracked oxide layer. EDS investigation of image b) shows that the carbon and oxygen content is slowly decreasing as a function of distance from the surface while the concentrations of the metals quickly approach the concentrations for the bulk, see figure 6. Carbon, oxygen and small traces of sulfur are abundant in the surface of the material. The chromium content close to the surface is reduced compared to the normal chromium content in the bulk material.

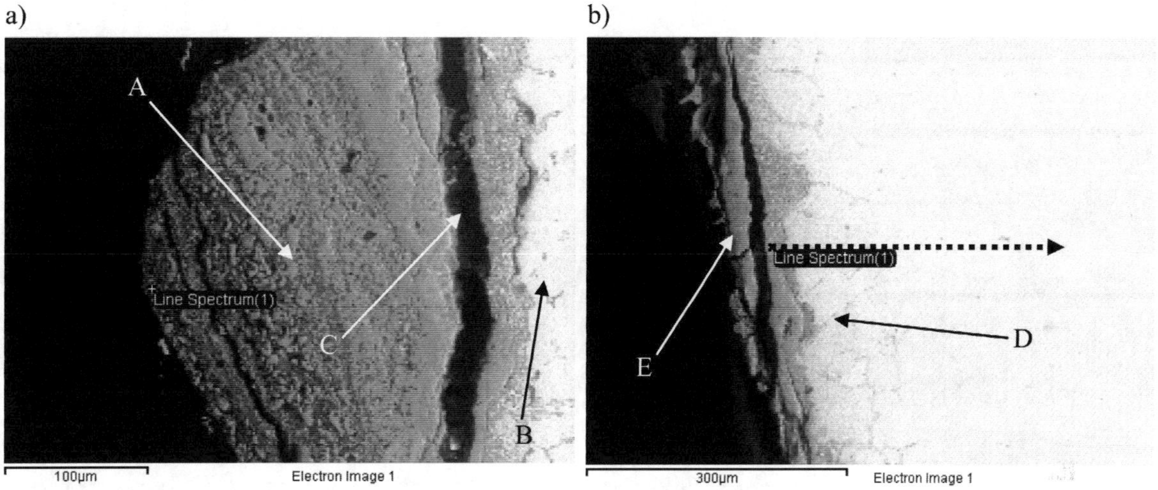

Figure 5. SEM images in BSE mode for the cross section of the oxide and the surface of the material. Deposits (A) can be seen on the surface of the metal (B) in picture a). A crack (C) is dividing the deposits from the metal. Thick grain boundaries (D) beneath a cracked oxide layer (E) can be seen in image b). The dashed arrows show the position, direction and length of the EDS analysis.

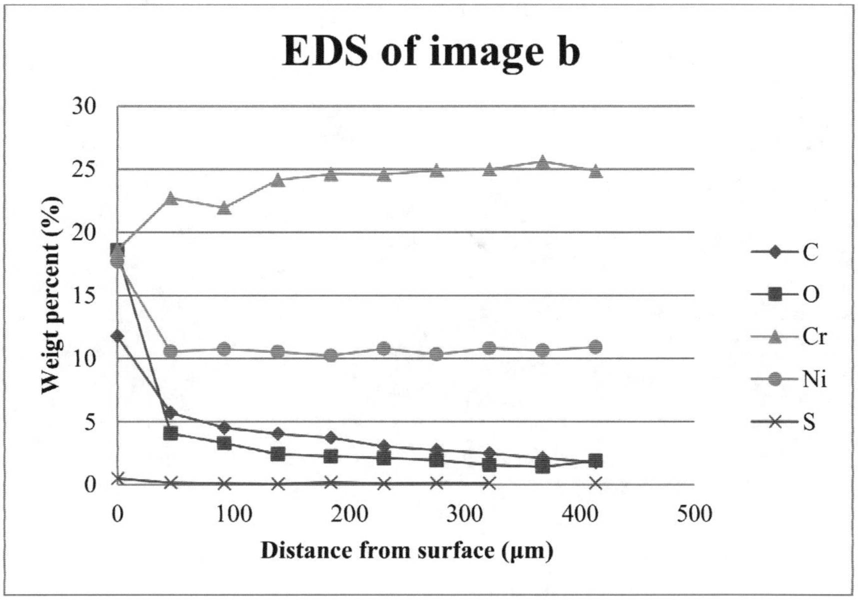

Figure 6. EDS measurements as a function of distance from the surface for image b) in figure 5.

3.2.3. Hardness measurements of rib cross section. Results from hardness measurements as a function of the distance into the material, together with standard deviations, are shown in figure 7. The hardness is decreasing as a function of the distance into the material.

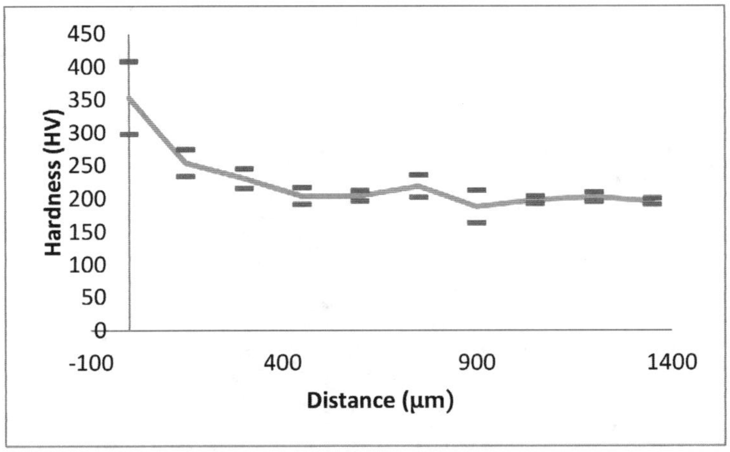

Figure 7. Hardness profile of the cross section of the ribs.

3.2.4. Microstructure of cross section. The cross section of the material was investigated by an optical microscope in dark-field imaging mode. Attacked grain boundaries are present as shown in figure 8 a) and b). The majority of the attacked grain boundaries are found at a distance up to 100µm beneath the surface. As can be seen in figure 8 b), there are some severely attacked grain boundaries that goes around 300µm below the metal surface.

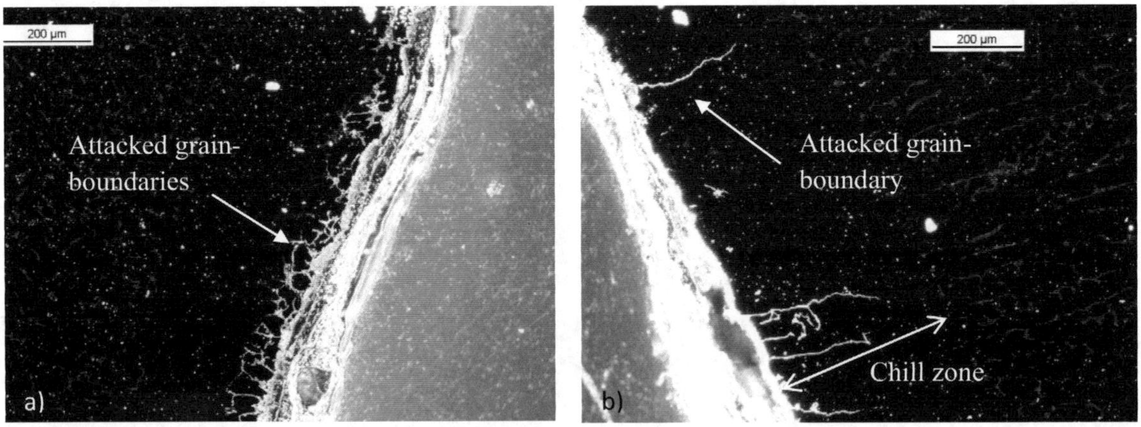

Figure 8. Dark-field imaging over the cross section of the surface of the material. The dark area between the surface and the dendritic structure represents the chill zone during casting.

3.3. Crack investigation of slit-edges

3.3.1. SEM and EDS-investigation. The crack propagation of the material can be seen in figure 9. The cracks have been following the grain boundaries which seem to be thickened in front of the crack tip. EDS analyses of the thick grain boundaries show that the carbon, oxygen and chromium content in these grain boundaries are higher than in a normal grain boundary.

EDS measurements of a crevice on the surface show that there is a significant difference in the chemical content in the inner end of the crevice compared to a point closer to the surface, which can be seen in figure 10. The inner end of the crevice contains high amounts of chromium and oxygen indicating a high amount of chromium oxide. The outer part of the crevice contains higher amounts of iron and oxygen which indicates iron oxide. EDS analyses also showed that chromium was depleted from the neighboring metal closest to the crevice. Such crevices are believed to be initiation points for crack propagation into the metal.

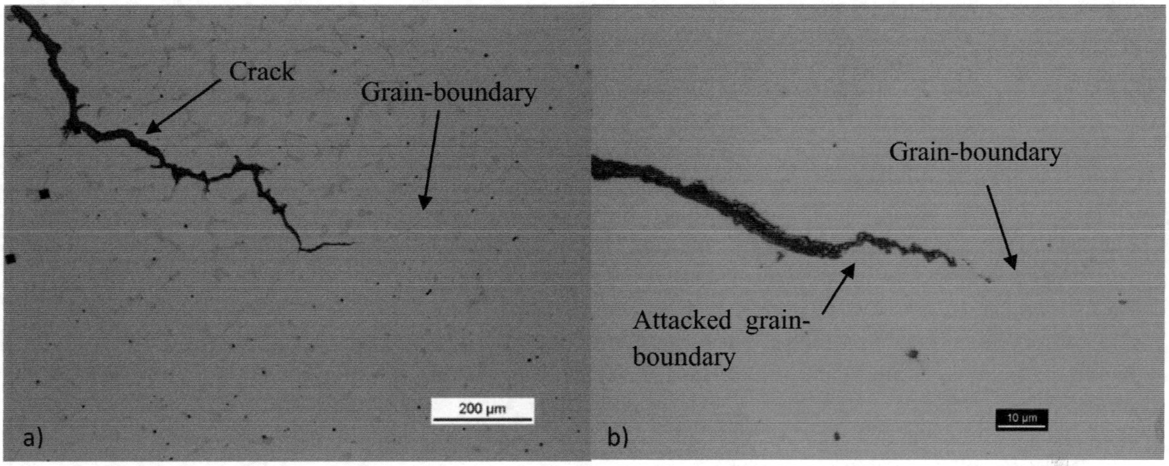

Figure 9. Crack propagation in the material. In a) a crack can be seen that follows a grain-boundary and in b) the crack tip can be seen to consist of an attacked grain-boundary before the actual crack arrives.

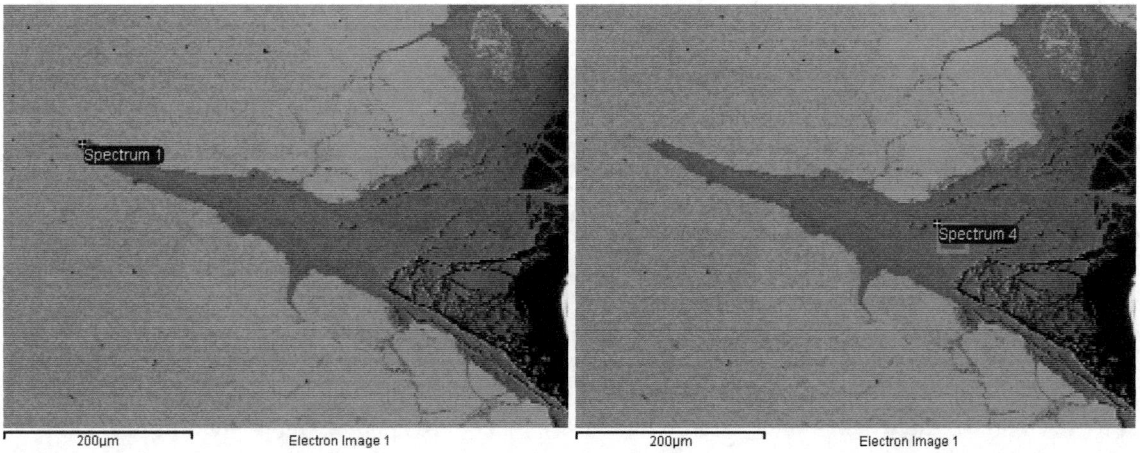

Element	Weight %
C	4,33
O	34,16
Cr	36,39
Fe	18,36
Ni	3,43

Element	Weight %
C	3,73
O	33,28
Cr	10,92
Fe	45,59
Ni	5,52

Figure 10. EDS measurements of two positions at a crevice, believed to be initiation points for cracking. The results of the EDS measurements can be seen below each image respectively.

IOP Conf. Series: Materials Science and Engineering **48** (2013) 012012 doi:10.1088/1757-899X/48/1/012012

4. Discussion

The inhomogeneity of the relative loss of the thickness of the ribs after exposure could be explained by the thermal differences between the measured positions. The center rib will have a higher temperature than those closer to the sides of the grate-link due to higher exposure rate to hot gas and smaller possibility to conduct the heat away from the rib to the bulk of the grate-link. This is also the reason why the measured positions closer to the slit-edge of the grate-link had a reduced loss of thickness for all the ribs. Another explanation could be that the gas flow is different at different locations on the grate-link which could be caused by clogging of the slits in the grate-link.

EDS shows that carbon, oxygen and sulfur contents decrease with distance from the surface and that carbon reaches normal quantities for the alloy 200 μm below the surface.

The hardness evaluation shows decreased hardness from the surface and inwards and the hardness reaches normal values of 220 HV [9] around 200 μm into the material. This suggests an inward diffusion of carbon which forms hard chromium carbides. Also the thickened grain boundaries close to the surface supports this mechanism.

Chromium oxide is a protective oxide layer and does not allow any high diffusion but the study has shown reduced amount of chromium at the surface which implies chromium depletion. There are generally two different mechanisms that could weaken the oxide layer in this case, a chemical or a mechanical mechanism.

Reduction of the effectiveness of the oxide layer by chemical means could be described by the complex chemistry of the environment with alkali metals, sulfur and water vapor. Literature suggests that the alkali metals Na and K with their sulfates, hydroxides and chlorides enhance the corrosion mechanism which is a process known as hot corrosion. These salts generally have a low nominal melting temperature and when they are mixed together their melting temperatures will be even lower. These melted salts dissolve and disperse the protective oxides so that they become porous and less protective. Hot corrosion resembles aqueous corrosion since a liquid increases the ionic transportation of the system [10, 11].

A second possible chemical mechanism that weakens the oxide layer is related to the role of water vapor in the environment since chromium oxide is unstable in hot environments containing high amounts of water vapor. Two different theories have been proposed to understand this mechanism. The first is a process of hydroxide ion uptake of the oxide which renders the oxide more prone to ion transportation which reduces the effectiveness of the protective layer. The second mechanism deals with the stability of chromium as a gaseous phase and studies have shown that chromium evaporates at 600 degrees in an environment with O_2 +10%H_2O(g) [12, 13][14].

Thermal cycling is a mechanism that could remove the protective layer mechanically. The difference between the highest and lowest temperature of the grate-link makes the protective oxide layer spall from the surface. This is due to the difference in linear thermal expansion of the oxide and the metal which causes stresses at the border between the metal and the oxide. The oxide layer also becomes more brittle when no melt is present as the temperature is decreasing which further increase of the rate of spallation [15].

Hot corrosion is described to have an initiation period where the alkali salts are accumulating on the surface and the material still resists the attack and it is reported that the oxides become thick and flaky after service [7, 16]. The ICP evaluation has confirmed alkali metals inside the cracks and in the oxide of a total of 5 percent.

While these theories explain the degradation of the protective layer they fail to explain the cracks inside the material and the thickened grain boundaries. Sensitization to Inter-Granular Attack (IGA) is a mechanism that arrives from the stability of chromium carbides. Carbon reacts with chromium and depletes the chromium in the material close to the grain boundaries which in turn becomes susceptible to IGA. Sensitization to IGA normally occurs at temperatures above 400 degrees to ensure a high diffusion of carbon but below 800 degrees because carbides are unstable above these temperatures [17].

The carburization is probably responsible for the sensitization which makes the grain boundaries susceptible to IGA. Thickened grain boundaries with an increased amount of carbon seem to support this idea.

Stress concentrations in the slit-edges of the grate-link are likely to form the cracks that are observed there. As the sensitization and IGA progresses into the material along the grain boundaries, the stress widens the grain boundaries and forms a crack. It is suggested to define the cracks as a mechanism of Inter Granular Stress Corrosion Cracking (IGSCC) [18].

5. Conclusions

The degradation of a grate-link in a grate-kiln indurator has been investigated.

The difference in rib thickness is due to the thermal difference, or the gas flow around the ribs.

Alkaline vapor from the warmer zones in the indurator are believed to condensate at the cool grate-link and form chlorides and sulfates together with sulfur and chlorine which continuously accumulate on the grate-link during service.

The alkaline salts cause hot corrosion on the grate-links. The H_2O (g), CO_2 and SO_2 content in the environment together with thermal spallation render the surface susceptible to carburization, sulfidation and internal oxidation. This could lead to sensitization towards IGA and thereby renders the material damaged by IGA. At places where stress concentrations are present it is proposed that IGSCC forms cracks in the grate-link.

6. Acknowledgements

This project is sponsored by the Hjalmar Lundbohm Research Centre (HLRC) and the mining company LKAB. Thanks also to Ragnar Tegman who has been a reliable support during the project.

7. References

[1] Lo KH 2009 *Materials science engineering.R, Reports* **65**:39
[2] Kain V 2011 *Eng Failure Anal* **18**:2028
[3] Tribouilloy L 2007 *Canadian Nuclear Society - 13th International Conference on Environmental Degradation of Materials in Nuclear Power Systems 2007* **2**:937
[4] Xiao J 2012 *Corrosion* **68**:015004
[5] Yoon KB 1999 *Eng Failure Anal* **6**:101
[6] Mohanty BP 1992 *Handbook of Case Histories in Failure Analysis* :358-362
[7] Mohanty BP 2004 *Corros Sci* **46**:2893
[8] Liu J-Y 2009 *International Journal of Minerals, Metallurgy and Materials* **16**:632
[9] Nilsson E 2011 *Degradering av grate-plattor i ett grate-kiln system*
[10] Kofstad P 1988 *High Temperature Corrosion*
[11] Lai GY 1990 *High Temperature Corrosion of Engineering Alloys*
[12] Pettersson J 2008 *Materials science forum* 595-598 **1**:367
[13] Rapp RA 2002 *Corros Sci* **44**:209
[14] Mäkipää M 2004 *Mater Sci Forum* 461-464:989
[15] Osgerby S Berriche-Bouhanek K and Evans HE 2005 *Mater Sci Eng A* 412:182
[16] Mohanty BP 2004 *Corros Sci* **46**:2893
[17] Easterling K 1992 *Introduction to the physical metallurgy of welding*
[18] Trillo EA 1995 *Materials Characterization* **35**:99

IOP Conf. Series: Materials Science and Engineering **48** (2013) 012013 doi:10.1088/1757-899X/48/1/012013

Effects of non-hydrostatic pressure on electrical resistance of bundled single-wall carbon nanotubes

M Noël[1], Y Volkova[2], M Mases[1], P Zelenovskiy[2], A Babushkin[2] and A V Soldatov[1,3]
[1]Department of Engineering Sciences and Mathematics, Luleå University of Technology, S-97187 Luleå, Sweden
[2]Institute of Natural Sciences, Department of Physics, Ural Federal University, 620002 Yekaterinburg, Russia
[3]Department of Physics, Harvard University, MA - 02138 Cambridge, USA

E-mail: Alexander.Soldatov@ltu.se,

Abstract. Recent studies have shown that single wall carbon nanotubes (SWCNT) exhibit a sequence of phase transitions and demonstrate a high structural stability up to 35 GPa of quasi-hydrostatic pressure [1] beyond which an irreversible structural transformation occurs. Here we report on the study of electrical resistance of SWCNTs at pressures up to 34 GPa in the temperature range of 293 – 395 K. In the pressure range 10-25 GPathe rate of resistance change decreases considerably. We associate such behavior of the resistance with a structural modification of the SWCNTs or/and change of the conductivity character at high pressure. Raman spectra of the samples recovered after 30 GPa exhibit a large increase of defect concentration in the CNTs. Isobaric temperature dependences of the CNT resistance R(T) measured in the temperature range 300-400 K reveal some changes with pressure whereas the semiconducting character of the R(T) remains unaltered.

1. Introduction

According to recent experimental data [1,2], single wall carbon nanotubes exhibit a sequence of structural transitions and demonstrate a high structural stability up to 35 GPa of quasi-hydrostatic pressure. The authors [1] tentatively relate an irreversible transformation of SWCNTs which occurs above 35 GPa to theoretically predicted polymerization of the material [3]. Change of CNTs structure at high pressure/stress may affect profoundly the electronic properties of nanotubes and even cause semiconducting to metal (and vice versa) transition [4]. Whereas several important Raman studies addressing the problem of SWCNTs [1,2] and DWCNTs [5, 6] integrity at high pressure have been recently reported they suffer from a common problem - only a fraction of the CNTs resonant with the laser excitations used in these experiments has been followed. This drawback can be eliminated by following evolution of CNTs' transport properties at extreme pressure as that will probe the entire nanotube ensemble. Importantly, no measurements of electrical transport properties of CNTs at high pressure/stress have been conducted to date. Here we report on the study of electrical resistance of SWCNTs at non-hydrostatic pressures up to 34 GPa in the temperature range of 300 – 400 K and correlate the results with correspondent structural changes of nanotubes.

IOP Conf. Series: Materials Science and Engineering **48** (2013) 012013 doi:10.1088/1757-899X/48/1/012013

2. Experimental methods

In this work we used purified CNTs which had been produced by HiPCO method by Tubes @ Rice Ltd. This type of nanotubes offers an advantage of a wide range of tube diameters typically ranging from 0.7 to 1.3 nm thus allowing probing a response of very different CNT species to the extreme pressure/stress. A certain amount of residual metal catalyst remains in the material as evidenced from the transmission electron microscopy (TEM) analysis (figure 1). Raman characterization of the material before and after high-pressure experiments was done on NTEGRA Spectra probe nanolaboratory confocal microscope from NT MDT, with He-Ne excitation laser (632 nm). The spectra were collected in back scattering geometry with resolution of 2.3 cm^{-1}. A 100x objective with a numeric aperture of 095 was used for focusing. Laser power was measured directly on the microscope sample stage and did not exceed 5mW. A typical Raman spectrum of the source material collected at ambient conditions (figure 1) displays the main features of CNTs, radial breathing mode (RBM) at low frequencies (from 150 to 300 cm^{-1}), D-band (characteristic of defects on the nanotube surface) and G-band at 1350 cm^{-1} and at 1600 cm^{-1}respectively. One should notice the relative low intensity of D band in comparison to G band which is indicative of low defect density in purified HiPCO CNTs used in this study.

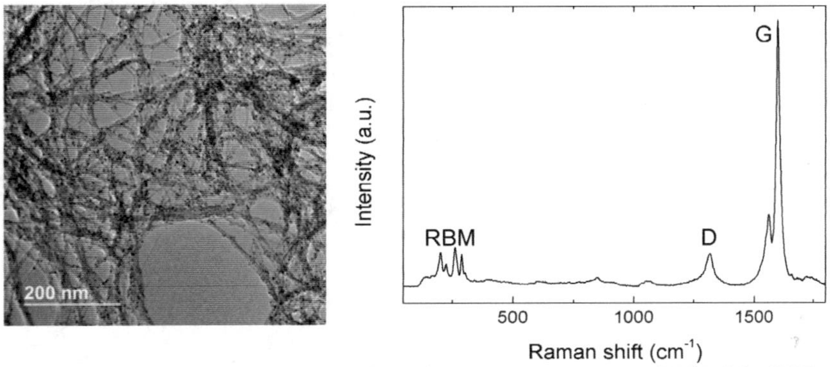

Figure 1.TEM image (left panel) and Raman spectrum of HiPCO CNTs collected at ambient conditions (right panel). RBM, D- and G-band are indicated.

High pressure was generated in a cone-plane cell (figure 2) with anvils made of carbonado type diamond [7,8] which allow for resistance measurements under pressures up to 50 GPa [9]. A cross section of the contact area between the anvils and the sample under investigation is typically around 200 μm. The sample thickness after pressure-induced compactization is approximately 10 μm. Pressure in the cell was determined from the applied load using calibration data for reference materials with error not exceeding 10% in the pressure range of 5-50 GPa [9]. First, the CNT sample was pressurized about 30 GPa with subsequent pressure release followed by a second pressure cycle which was conducted in order to examine reversibility of the R(p) dependence. CNT resistance dependence on temperature R(T) was measured between 300 and 400 K at two selected pressures (20 and 30 GPa). Temperature uncertainty was about 0.1 K

IOP Conf. Series: Materials Science and Engineering **48** (2013) 012013 doi:10.1088/1757-899X/48/1/012013

Figure 2. The conductive (carbonado) diamond anvils (left panel) and schematic cross-section of the plane – cone pressure cell (right panel), not to scale

3. Results and discussion

3.1. Pressure dependence of electrical resistance of SWCNTs

The electrical resistance decreases as pressure is increased from about 80 Ohm at 12 GPa to less than 30 Ohms at 28 GPa for the first pressure cycle (figure 3), The resistance evolution with pressure shows a first anomaly at approximately 16 GPa and another one at around 25 GPa. This first irregularity appears at the pressure corresponding to the collapse pressure Pc of the CNTs with the largest diameter (1.3 nm). In addition, the second change in the pressure dependence of electrical resistance is also related to the structural change of the large tubes, as 25 GPa corresponds to the collapse pressure of tubes with a diameter of 1.1 nm.

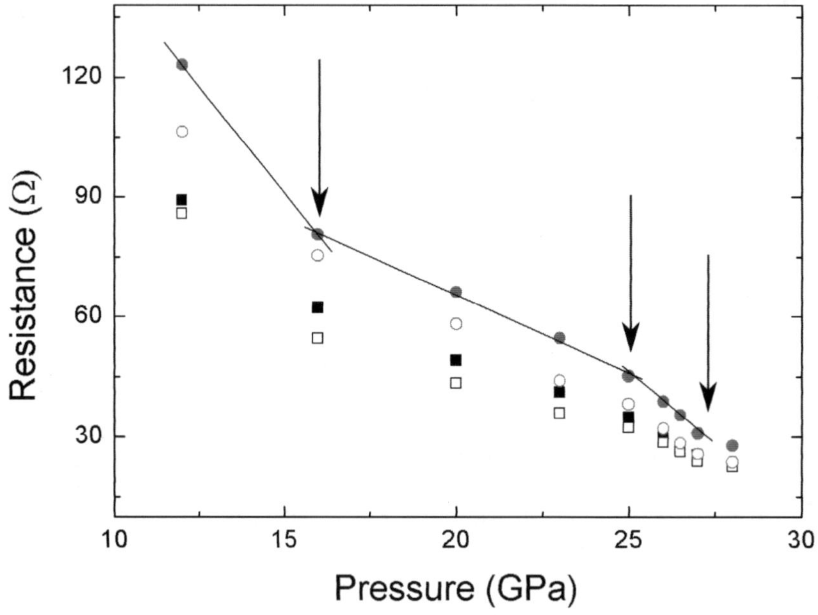

Figure 3. Pressure dependence of the resistance of the SWCNTs nanotubes for the cycle 1 (squares) and cycle 2 (circles). Solid and open symbols represent data collected on pressure increase and release respectively. Straight lines indicate quasi-linear character of the R(p), see text.

Theoretical calculations predict collapse of CNTs in bundles at high pressure [8]. The collapse pressure P_c depends on the CNT diameter d as $P_c = k/d^3$ where k is a constant. Experimental studies [2,8] report on CNT collapse at high pressure in accordance to the theoretically predicted dependence on nanotubes' diameter. Using the data on the SWCNT collapse [2,8] we estimated the collapse pressures range for SWCNT used in our study to be 16 - 100 GPa in accordance to the above mentioned formula for P_c with the highest P_c corresponding to CNTs with the smallest diameter (0.7 nm). The pressure range of our experiment falls into this interval and therefore it is reasonable to associate the peculiarities in R(p) dependence with sequential collapse of the nanotubes at correspondent P_cs on pressure increase. From the latter we can define diameters of the nanotubes which undergo this structural transformation. Our estimation gives CNT diameters of 1.3, 1.1 and 1.08 nm for P_cs 16, 25, and 27 GPa respectively. A 10% hysteresis in R(p) observed upon pressure release (figure 3) is probably connected to delay in the CNT shape recovery after collapse.

3.2. Temperature dependence of electrical resistance of SWCNTs at different pressures

Carbon nanotubes can be metallic or semiconducting depending on their chirality. In order to examine the overall electronic properties of the CNT material we performed isobaric measurements of the sample resistance in temperature interval 295 – 400 K at 20 and 30 GPa selected from the pressure interval before and after P_c = 25 GPa respectively. The observed temperature dependences of the resistance R(T) exhibit similar qualitative behavior (figure 4) – the resistance increase with decreasing T typical of semi-conducting material, implying that more than 30% of the CNT species in the sample are semiconducting. The R(T) dependencies are non-linear in ln(R) – 1/T coordinates below 370 K implying Ahrenius mechanism of electrical conductance does not hold in the entire T-range of our experiment (right panel of figure 4). Importantly, the character of R(T) at 30 GPa differs from that at 20 GPa. As certain fraction of CNTs undergo pressure-induced collapse at P_c = 25 GPa (figure 3) it is reasonable to associate altering the overall R(T) dependence with electrical conduction mechanism and gap change in the collapsed (flattened) CNTs. We even can not exclude the occurrence of semiconductor to metal transition in some CNTs caused by stress [4] although such transition is not dominant in the studied pressure range and thus can not be identified on the background of rest of the CNT ensemble.

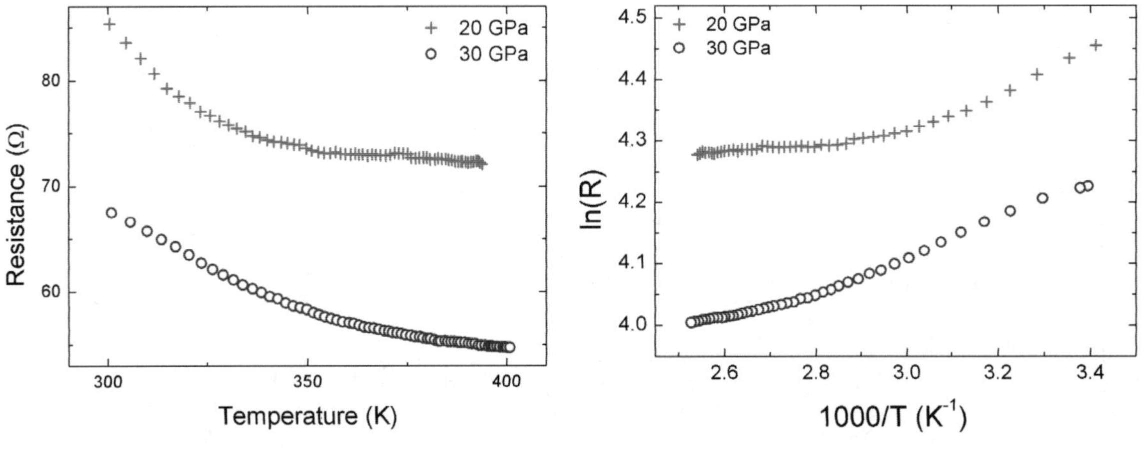

Figure 4. Temperature dependences of the SWCNT resistance R at selected pressures (left panel) and lnR vs 1/T plot (right panel). Crosses and circles – data for 20 and 30 GPa respectively.

3.3. Spectroscopic characterization of SWCNTs recovered from high-pressure

SWCNTs exposure to high pressure resulted in dramatic changes in their Raman spectra (figure 5). Specifically, the D-band to G-band peak intensity ration (I_D/I_G) has increased from 0.6 to 1 implying introduction of more structural defects on the CNTs surface. This result implies that a substantial fraction of the nanotubes has undergone collapse during the pressure cycling as it was shown in [6] the generation of new structural defects on nanotubes surface occurs when they are in a collapsed state. It is important to underscore that nevertheless, the RBM band remains essentially unaltered: the main peaks at 200, 252, 261, 288 and 301 cm^{-1} are still visible and display the same shape and relative intensity after two consecutive pressure cycles to 30 GPa. That testifies for the CNTs full recovery after the pressure cycling – neither irreversible structural transformation, nor polymerization of the tubes has occurred.

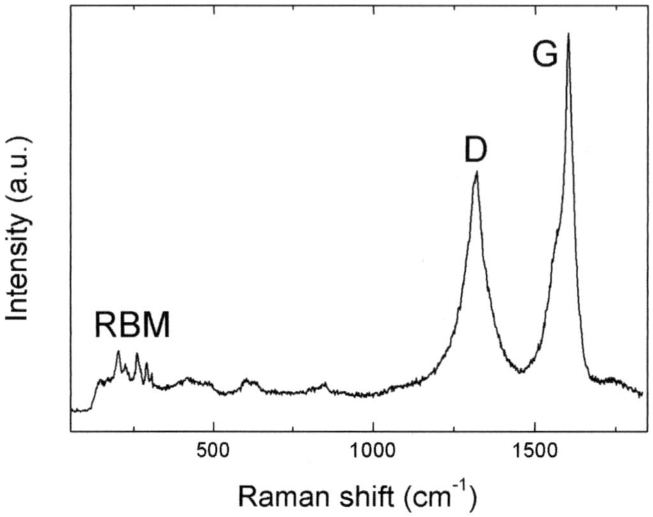

Figure 5. Raman spectrum of HiPCO nanotubes recovered after pressure cycling to 30 GPa. RBM, D- and G-band are indicated.

4. Conclusions

Peculiarities in pressure dependence of the resistance of HiPCO SWCNTs were observed at certain pressures which we relate to a structural transition – sequential collapse of the nanotubes. The collapse pressures are defined by CNT diameters. This transformation is accompanied by the I_D/I_G ratio increase in the Raman spectrum of the sample recovered after exposure to 30 GPa of non-hydrostatic pressure which is indicative of a high density of structural defects generated in the CNTs which have undergone collapse during the pressure cycling. The RBM band recovery implies a complete reversibility of the structural transitions and no CNT polymerization in the studied pressure range. Further experiments are required to ascertain the observed character of R(T) and probe structural stability of small (below 1 nm) diameter SWCNTs.

Acknowledgments

A V. Soldatov acknowledges a financial grant from the Swedish Institute. A. Babushkin acknowledges a financial support from the Russian Minister of Education and Science and Ural Federal University development program. M. Noël gratefully acknowledges J.C. Kempe & Seth M Kempe Minne foundations for a PhD fellowship.

References

[1] Chen J Y, Kim M and Yoo C S 2009 *Chem. Phys.Lett.* **479** 91–94
[2] Caillier C, Machon D, San-Miguel A, Arenal G, Cardon H, Kalbac M, Zukalova V and Kavan L 2008 *Phys. Rev.* B **77** 125418-125426
[3] Sakurai M and Saito S 2011 *Physica* E **43** 673–676
[4] Yang L and Han J 2000 *Phys. Rev. Lett.* **85** 154-157
[5] Aguiar A L, Barros E B, Capaz R B, Filho A G S, Freire P T C, Filho J.M and Machon D 2011 *J. Phys. Chem.* C **115** 5378-5384

[6] You S, Mases M, Dobryden I, Green A A, Hersam M C, and Soldatov A V 2011 *High Pres. Res.* **31** 186-190
[7] Verechagin L F, Yakovlev E N, Stepanov G N and Vinogradov B V 1972 *JETF Lett.* **16** 240-242
[8] Verechagin L F, Yakovlev E N, Vinogradov B V, Stepanov G N, Bibaev K K, Alaeva T J and Sakun V P 1974 *High temp., High press.* **6** 99-105
[9] Babushkin A 1992 *High Pres. Res.* **6** 349-356
[10] Elliot J A, Sandler J K W, Windle A H, Young R J and Schaffer M S P 2004 *Phys. Rev. Lett.* **92** 095501-095503

Institute of Physics
Dirac House, Temple Back
Bristol BS1 6BE UK

CURRAN ASSOCIATES INC
proceedings.com